Dark Matter from Light

extending quantum mechanics to Newton's First Law

by

John P. Wallace and Michael J. Wallace

2011

Casting Analysis Corp.

Casting Analysis Corp.
Weyers Cave, Virginia
www.castinganalysis.com

Part No. pm5-manual-vol-4-1.04

ISBN 978-0-615-51839-8

dedicated to the memory of

John Robert Wallace

and

Steven Thomas Wallace

Preface

An early outline of the work covered in this book was added as appendix C to *Proton in SRF Niobium* [1], which was used as a text for a short course on hydrogen's effect on niobium used in accelerator components sponsored by ISOHIM following the SSTN10 meeting on ingot niobium for manufacturing of resonant superconducting accelerator cavities at the Newport News Accelerator Facility on 24 Sept. 2010. It was decided to include the short course text into the conference proceedings as a guide for using induction techniques applied to magnetic materials.

In the afternoon session of the short course some time was spent discussing low mass particles, because one of Newport News' experimental programs was an Axion search. At the workshop luncheon, the low mass magnetic excitation in steel was named the pAxion using a convention of P.G. Wodehouse. Naming the excitation was useful, because giving it an identity precipitated a second look at an effective mass theory for very light particles that had consumed some time over the previous two years. In the middle of October, on checking the calculations in the short course text for allowed proton eigenvalues, it became evident that there were additional solutions to the spherical flat bottom potential problem that had been historically overlooked. It was apparent these solutions could be used to define mass for a boson. In the following two weeks, we were able to finish the analysis for boson mass. Just after this work was completed the pictures of the Milky Way energy bubbles from a NASA-Harvard-Smithsonian collaboration were published in early November. Data from the image allowed us to compute the photon's mass as an example application of the mechanism that produces mass.

From the time the mass of the photon was determined, a search began for a derivation that produces an equation that self-consistently captures mass. When it was realized the entire exercise was one of extending quantum mechanics to encompass Newton's first law into a basic equation of quantum mechanics the problem was solved. As a bonus, an explicit analytical definition of the rest state resulted and some new roles for the photon emerged.

John P. Wallace
Weyers Cave, Virginia

Michael J. Wallace
Draper, Utah

10 August 2011

Acknowledgments

This work began at a small nondestructive testing firm Magnetic Analysis Corp., where it was learned that magnetic materials did not always obey classical electrodynamics, particularly hot rolled steels. Information was gathered from many customer applications and led to the design of instruments to specifically investigate these properties. Geophysical field measurements provided more data and forced the instruments to operate over a wider dynamic range. A parallel thread from the beginning of Casting Analysis was the problem of using standard computer languages to acquire and model data in real time on multiple processors which was initially not simple until the Forth computer language became available and allowed a single person to maintain the necessary code, so it could run over generations of a varying collection of operating systems and hardware [2]. This generated a stream of calibrated material data that kept these problems in view.

A fellow Forth language programmer, Julian V. Noble, [3], who when we were stuck on electromagnetic induction problem in steel just before he died, suggested a look at the problems with mass and gravity [4] because that was a blind alley much like our EM problem. The electromagnetic problem was then treated strictly as a fields problem, which led to some key experiments that found a modest exciton mass defined on a macroscopic scale [5]. John David Jackson and Phillip Kim were kind enough to help in getting that experimental electromagnetic work published. Then Ganapati Rao Myneni's determination to find an understanding of hydrogen in superconducting niobium gave us the opportunity to extend our recent experimental work on hydrogen in iron [6] to niobium, whose data analysis led to a description of proton diffusion, that yielded a path to the mechanisms that defines mass. We want to thank John David Jackson for pointing out some of our errors in the use of the spherical Bessel functions and John S. Klinger for his comments. Editorial assistance from Jerome Dunn, Robert Pike and Eric Gorton was a great help in completing this book. None of this would be possible without the patience of Janice Wallace, whose help allowed Casting Analysis Corp. to spend some effort on research.

Contents

7 Comment on Quantum Mechanics   121

8 Appendix   125

Chapter 1

Introduction

The impact of real economic behavior on the efficiency of research is a topic that is of general interest to most engineers. A great deal of effort is expended on exploring the historical connections controlling progress [7], and we have noticed that most practicing engineers are curious about the history of their professions. As the number of participants grows, the economic and social processes described by Adam Smith [8] come into play and in particular, is the division of labor into specialized practices. This economically favorable process by itself can tend to quench innovation because it cuts off access to original disciplines, which can wither without new perceived problems to solve from within, and being isolated develops resistance to external inputs. This fracturing of the disciplines is only balanced by a weak flow of information across boundaries, that seems to operate more on a generational time scale than an electronic time scale of the current age. Eighteenth and nineteenth century engineering and science was largely self funded by amateurs and really only persists in astronomy and some areas of biology. Its not a bad model because the investigator is probably more wisely going to allocate his own resources and it is free of the distractions of an organizational culture that biases directed research efforts. It is a model that rarely allows for support of elaborate research tools such as accelerators or satellite telescope platforms. But it is a model that is still useful when looking at problems that cross these naturally formed divides. This book is a by product of looking at two very mundane problems; finding flaws in hot rolled steel and explaining the source of anomalous magnetic properties in the cryogenic plumbing of particle accelerators [9]. The work on steel, which covered a number of years, resulted in finding a magnetic excitation that had an effective mass of nearly one billionth that of an electron. This was one of those things that was not expected, but stirred interest in the origin of low mass objects. The problem with the superconducting-pipes, which form resonant RF cavities, are usually only seen when the accelerators are run at their performance limits. Niobium is the best single component metallic superconductor. The cavities which the pipes make up can be driven into resonance with modest microwave input power and drive oscillating electric fields as has high as 50 million volts per meter. The manufactured cavities get contaminated with atomic hydrogen which is absorbed readily from the acids used to clean damage from its interior surfaces. Since it is absorbed in parts per million concentration, it has been largely ignored. Similar things happen with hydrogen in steels, where the effects are more dramatic and have been studied for many decades with little resulting understanding. However, to optimize the niobium accelerator a detailed understanding of hydrogen within the metal is required. The practical reason to improve these accelerators is to operate at high beam currents to generate a copious and economic source of neutrons than can run controlled sub-critical nuclear reactions producing medical isotopes, energy and driving transmutation of long lived nuclear waste.

The approaches used to solve this problem involves bringing in experts from three disciplines: physics, metallurgy, and chemistry along with the manufacturing details to approach the problem. But in fact each of those disciplines will supply two to four sub-discipline experts. Now you have selected a group that is designed not to communicate very efficiently because they are in effect speaking different languages. This approach has failed in the case of the accelerator and in many other cases of similar complexity. Each discipline has so diverged from the others to such an extent there is very little common ground and a lack of self confidence in each parties ability to cross over and discuss the problem with the other disciplines. The reason is that the individual parties are trained by learning a set of common problems for their discipline, which they have generalized into a set of rules for looking at problems. If the common problems the individual specialist were trained on have little or no meaning for the current problem being investigated then there is no common vocabulary with which to interpret data. Also not knowing the common background of the other disciplines makes asking questions that could possibly be answered, difficult. For the politically minded researcher this type of problem is ideal since it has no apparent solutions and can be used as a source of funding indefinitely, or as long as one is able to stop progress either consciously or unconsciously.

In order to make progress on this plumbing problem, it helps to revert to the model of the amateur researcher and look at each discipline to extract what is useful for the problem. At the beginning you cannot be very selective because, the important information and questions are yet unknown. More importantly you don't know what may be wrong or poorly understood with the discipline you are trying to learn about. It sometimes helps if you have learned a discipline and then forgotten it and you have to relearn it again. You will probably be more skeptical the second time, asking questions and seeing weaknesses in the discipline's development. This is exactly what was done for the case of the dissolved proton in the pipes. After surveying the metallurgy of defects generated by hydrogen, the structural chemistry of the oxide and electro chemistry of possible catalysis paths to drive hydrogen into the metal, we came to consider the physics of atomic hydrogen in metals as being important to understand the superconducting properties. But more importantly we had been taking data on the magnetic properties of the hydrogen contaminated plumbing. This gave us a working familiarity with the metal in a hands on fashion. Radio frequency magnetic measurements give you a measure of the dynamics of electronic ordering than can occur with very small perturbations of the lattice caused by the hydrogen content. Weak effects not strong effects that disturb equilibrium are often a great deal easier to understand. It was also a tool that was actually sensitive to these part per million quantities hidden in the metal that are difficult to study. The density function theory and perturbation approaches which physics normally uses to treat a dissolved ion in metals fail in this particular case, because the scale on which the problem must be solved is below a critical threshold for these techniques. This is where the amateur is at an advantage, he is free to select the best tools available rather than forcing a favored tool of a discipline where its not ideal. So a decision was made based on a long screening length and the magnetic data to treat the hydrogen as a dissociated proton and free electron as the starting point.

In the development of any model the critical act is the selection of the starting assumptions, because these will be difficult to overcome if they are wrong. Solving the quantum mechanical problem of the proton in the lattice is the first step. That turns out to be one of the simplest problems in quantum mechanics and because of having forgotten most of the subject a second look turns up a difficulty. In fact in the classic texts, it is often the first problem solved for students. So all of the experts will think they know the solution at least in one dimension, which is a trap because the problems of interest are three dimensional. The problem was solved and everything was fine. The solution produced data on the proton that allowed a direct computation of the activation of energy for diffusion for the proton and its isotopes, which were accurate. This worked not only for hydrogen in niobium but for hydrogen in a few other metals. We now have an understanding of the mechanism of rapid diffusion of hydrogen in metals. However, on our last check of the calculation we noticed there were more solutions that had been ignored.

Actually, these new solutions had probably been ignored for about 80 years. These solutions were different and not particularly interesting for the problem of the proton in metals. Immediately upon their appearance, it was realized they provided a path to understand mass. The way this problem is set up, it is the natural approximation for a particle which is attractively interacting with space in which it exists. The form of the problem gets around one major obstacle in doing quantum calculations where many particles are involved, a cumbersome method of calculation called perturbation theory, and provide some direct insight into the concept of mass. Taking advantage of these new solutions to simplify a many-body calculation was too great a temptation to ignore. What follows is that same approach used by the amateurs on the niobium plumbing, first trying to gain a microscopic understanding of mass from what we considered first principles, but then getting lucky again by finding a way to bring quantum mechanics up to date with Newton's first law. We have pushed the amateur science model into astronomy and tried to see what some of the implications are for the photon. We could have just as well pushed the model into finance, economics or sociology because this new description carries an extrinsic definition of value that is coupled to a very elementary formalism. This introduction started with a reference to the problems in the history of innovation and ended with a method for a simple approach to research across different areas. It is not apparent that the forces described by Adam Smith that drive the differentiation of labor are all that helpful when working on problems where the science is partially understood. This feature of labor economics gives the amateur a natural advantage. When the approach is applied it yields a random walk that is simply trying to avoid the traps laid by individual disciplines.

1.1   The Mass Problem

1.1.1   History

Of all the physical things we can measure, mass has one of the thinnest bibliographies in the physics literature [10]. Mass as a measure is a positive scaler quantity that is made up of the sum of isolated individual components. It transforms inversely as compared to a scale length in special relativity , thereby gaining in value as its velocity is increased. It also has a linear proportionality to total self energy. Experimentally, it became interesting when Galileo dropped his two different masses demonstrating the uniform quality of the property independent of chemistry or quantity [11]. Analytically, mass first made its appearance in Newton's third law as the coefficient that connects a force to acceleration. Newton extended the use of the concept by making mass the source of a particular force, that of gravity with the $\frac{1}{r^2}$ force law [12].

These findings then allowed Ernst Mach to propose his principle, that led Einstein to the equivalence principle, which states that one cannot discern the difference between the field of a gravitating body or an applied acceleration if they are of the same magnitude. Also, it has been shown that for a charge carrier like an electron, its inertial mass is not solely a feature of its electromagnetic field [13]. Unlike the electromagnetic interaction, where the carrier of the force is an electromagnetic field that can freely propagate in space, no such freely propagating field has been found for transmitting the force of gravity. Einstein in the general theory of relativity avoided this problem by endowing mass with the property to deform space and make the local curvature proportional to the local mass. The general theory of relativity is an accurate macroscopic theory explaining the dynamics of matter and space, which implies there must be a microscopic theory of mass. There are a number of quantum gravity theories, but in a recent popular review of the subject, mass itself does not even make an index entry. Actually, these theories would best be described as submicroscopic and to slow this rapid decent to the Planck scale of 10^{-35}m, intermediate theories have been explored [14] that look at what is required to be consistent with new experimental data on the behavior of wave functions in multi-particle systems. This short paragraph reviews what is known. The difficulty has been that there are no known connections that allow mass to be described in terms of other physical parameters other than energy.

One area that has an extensive bibliography is the question: Does the photon have a rest mass? This area has been well investigated with strong opinions by many parties. If a photon is found to possess a mass, it changes the way a number of calculations are performed and has a major impact in astrophysics and cosmology. Experimentally, only the upper bound on the photon mass has been established, and this subject is reviewed at the beginning of *Classical Electrodynamics* by J.D. Jackson [15]. The answer is rather important because of the way the electromagnetic forces are understood to be mediated by the transfer of photons. The local properties of electromagnetic fields show that the photon mass is zero. However, the large scale properties of photons affecting processes on astronomical scales may show unexpected characteristics. Prior to starting those inquiries, a better understanding of what mass actually is and how it can be defined is necessary.

The first introduction to the question of what is mass and how it fits in with mechanics and dynamics was raised in a freshman physics course. Because the professor lacked a current familiarity with classical mechanics, he thought it was better to start with quantum mechanics, a much more comfortable subject. While introducing the two standard solutions to the Schrödinger equation, the harmonic oscillator and the hydrogen atom, some questions were raised. A mathematics student named Peter Landesman asked a question in the late fall of 1966 to the professor Jack Steinberger about mass. The question, what were the implications in solutions to the Schrödinger equation if mass were a complex quantity? The instructor said he did not know, but said he would ask someone. The next lecture the instructor announced that a complex mass would lead to some kind of decay process. A few months later Gerald Feinberg came out with his proposal for a faster-than-light particle, the tachyon [16]. The guess is that the freshman's question started the process. What makes the Schrödinger equation different from the diffusion equation is the fact that it explicitly contains the complex quantity, i. $i \times i = -1$ is the arithmetic identity, but the appearance of i early on in 1673 was used to construct a two dimensional space because there was no simple equivalence of real and complex numbers by Wallis [17]. To introduce a second complex quantity, via the mass factor, returns the free particle version of the Schrödinger equation to one familiar in classical mechanics, and the quantum solutions are lost. The approach of making mass complex did not seem to be very interesting at the time, but it did leave a further question unanswered. Why should mass be limited to positive scaler values?

Twentieth century physics and quantum mechanics originated from a very fruitful study of thermodynamics and statistical mechanics [18]. When an algebra was created that gave a probabilistic basis for measurement, the principle features of quantum mechanics emerged. This was possible because the problem separated into an analytical hierarchy. The governing equations produced solutions where the energy and momentum values were real quantities, and the wave functions were complex. The probability densities defining the quantum particles distribution, which were measurable, were the wave function's norms, and they were also real. The development of quantum mechanics into quantum field theory appeared to take on a more analytical aspect as if trying to minimize its origins from statistical mechanics even though it was enumerating many-body interactions in a formal manner. The features that remained were the required algebra for the behavior of the wave functions, which made the connections to experiments where they could be verified. Unfortunately, in the development of the subject one of the aims was to derive quantum mechanics and quantum field theories from a classical model of mechanics using either the Hamiltonian or the Lagrangian approach. This was not entirely successful because the basis of quantum mechanics was statistical. These classical approaches started with a single particle and field as the principle element. It is not sufficient to model these problems classically and then by analogy transfer them to a quantum format, which resulted in grafting a theory rather than deriving the theory. Building a theory from the experimental algebra is the favored approach, but it still falls short of naturally producing the fundamental equations. This is probably the reason that the fundamentals and foundations of quantum mechanics are still an active area where experimental data is exposing new ground. The problem of how mass arises became an interest when we stumbled [5] on macroscopic fields that had effective masses much smaller than anything previously measured. It became apparent that the method to derive the Schrödinger equation must also provide a description of mass. The coupling between mass and the Schrödinger equation results in a microscopy theory or intermediate scale theory of mass, and it should provide some basic information about the connection of mass to the general theory of relativity.

Summary

What follows is a look at the foundation of quantum mechanics by a metallurgist and a geophysicist. It is a look that is going to stress, as in deform, a working quantum model in a non traditional way. We will just deal with one simple point, the self consistency of the definition of mass in a quantum frame work. The basic points of the discussion are not complex. Even the forms of the new solutions presented which may look formidable are very simple physical solutions and well behaved. The concepts can be grasped completely by a interested amateur. On the subject of the astrophysical implications, only one of us has worked in the area for a short time. However, these discussions are so important to the understanding of mass that it trumps the possibility of making some embarrassing mistakes of omission or ignorance.

1.1.2   Elementary Solutions

The two most elementary solutions in quantum mechanics are the hydrogen atom and the harmonic oscillators, which are the basis of much that we know about the allowed responses of a single particle and a potential. In trying to understand the ground state behavior of protons in metals [19], a third model system at a basic level capable of modeling a many-body interaction with a single particle appears to exist. The major features of quantum mechanics originally came from solving just two problems. These solutions were extended to the Cartesian flat bottom potential problem for modeling electrons in metals for solid state physics. The simple model that was not adequately explored is the flat bottom spherical potential problem. None of the three dimensional elementary quantum mechanical closed form solutions are analytically simple except for the solid state potential problem. They provide a more direct means of exploring the individual problems. In particular, some overlooked solutions of the spherical potential problem will be useful in providing insights about the origin of mass and the Schrödinger equation itself. The problem is initially defined as a unit potential of finite radius and spherical symmetry that can either be positive or negative. It is taken as an elastic unit potential that can be deformed to its limits or even transformed to a hybrid of the two interactions to locate possible eigenstates. The spherically symmetric flat bottom potential differs from the $\frac{1}{r}$ Coulomb potential and the x^2 potential in the harmonic oscillator because there are two free parameters, not just the magnitude of the potential. For the flat bottom potential, the two free parameters are the magnitude of the potential and its dimension. The dimension is introduced explicitly, which is not done either for the Coulomb or for the harmonic oscillator potential problems. The potentials are sketched in one dimensional cuts, their three dimensional structure cannot be shown in a two dimensional illustration, but can be easily imagined.

Figure 1.1: **Simple Potentials.**

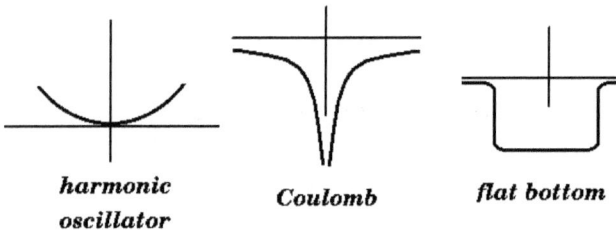

harmonic   **Coulomb**   **flat bottom**
oscillator

Selecting a simple potential to describe the background interaction with vacuum or in the solid state does not require one initially to identify and enumerate the interactions involved. In fact, the potential itself is not even considered a fixed uniform potential, but one that represents an average of the activity involved. The sum of the interactions are simply lumped in the potential. The interesting feature of this potential is that in the weak limit the attractive potential produces only three allowed eigenstates for angular momentum: $l = 0, 1$ *and* 2. And in the repulsive case though there are two solutions, only one interesting solution at all scales for $l = 0$ will probably be important. The immediate implication from the dependency of the parameters forming the eigenvalue is a connection to the definition of mass in the limit of the weakest interactions.

To explore the origins of mass, the most promising place to start is a quantum mechanical model of the particle interacting with its space. This approach is a little more elementary than a direct approach taken by Wentzel for the photon [20] and more general because of fewer conditions than the approach taken on nuclear matter [21], [22], [23], [24], which imposes a requirement of a particular boson field modeled on a harmonic oscillator with a singularity for a frequency greater than zero producing mass. The method we wish to use makes few assumptions while retaining enough of a problem to make progress. It would be convenient to examine the interaction in the frame of reference of the moving particle, but this can only be an approximation because the random interactions modeled by the potential makes selection of the frame of reference an approximation. As a starting point, we will assume the particle interactions with the space are weak. Secondly, in the particle's frame of reference, the particle's density integral over space is required to be finite. This defines a hierarchy in the analysis of a particle's behavior by separating the dynamics of the particle interacting with external potentials from that of its interaction with the vacuum state. This separation in the description of the particle's properties will be reflected in the resultant model.

If we look at the particle's weak interaction with the vacuum state in a model using the flat bottom potential describing the polarization, these conditions should only produce a small contribution. In particular, if the definition of mass is entangled with the derivation of the Schrödinger equation itself, then those details must be understood. We will start with the hypothesis: *mass is a quantum mechanically derivable property that must come from a solution of the particle interaction with space to yield a small finite solution set.* This hypothesis is based on the concept that inertial mass presents a resistance to action characteristic of particle type. This resistance to action is due to the particles base interaction with the space it occupies. One good example is the finite set of quantum properties of spin which comes with two solutions [25]. Mass should be simpler since it comes with just one solution per particle. Mass has no detailed classical definition, only how it obeys dynamic and gravitational laws. If mass has its origins as a quantum property, then a strict application of the correspondence principle where quantum properties go into classical properties can be applied. If the quantum numbers are restricted to one or two, then the properties are the same in classical and quantum applications.

The motivation for this interest in mass, and particularly low mass entities, is because of recently finding a magnetic excitation in iron and steel [5] having an effective mass $\sim 10^{-9}$ less than the electron's mass. We call these excitations pAxions for convenience. In the solid state, to understand mass, one must consider the many-body interaction a particle encounters. The potentials generated in specific instances may not be uniform, but random in character, and this is an important limitation on our detailed knowledge of the process. To keep things general at the start, no specific potential sources need be considered until the results are applied to a specific particle. The interaction of a particle with its medium can have only three outcomes:

- No interaction

- Particle polarizes[1] the local medium and it is attracted to the polarized region

- Particle polarizes the local medium and it is repelled from the polarized region.

There is also the implied condition that we are looking at the properties of an elementary particle, not a composite particle such as

---

[1]The word polarize here means nothing more than to affect. It has no connection to the very specific plasma oscillation model of Nozière and Pines [26] generalized by P. Anderson [23] and others to provide a mechanism for the origin of mass. The model of Nozière and Pines requires two oppositely charge distributions of matter, where one component is much more massive than the other. These are models appropriate for high density and very energetic multicomponent applications where existing collection of masses are required for the excitation to exist.

a nucleon. The trial potential that will be used to find the conditions for mass will be a spherically symmetric flat bottom potential that is either repulsive or attractive. This is a useful potential because it behaves as a simple unit that has two useful parameters, depth and radius. If the potential is scalable algebraically, it can reproduce a more complex potential. Since this is a search for the conditions of mass, the results require this potential to be scaled so that the final resulting potential may be more complex.

The first information that will be searched for in these last two cases is what are the specific conditions required for a particle to have mass? Secondly, is there anything fundamentally different between the effects of an attractive and repulsive potential? This information alone, if it were unambiguous, would be helpful. If the results of these two model problems are interesting, then narrowing down the potential sources in the vacuum state for stable particles and within iron for the pAxion would be necessary to complete a physical description.

The spherical flat bottom potential is an important potential because it is the simplest potential that can represent a many-body interaction in a compact manner. Both the $\frac{1}{r}$ and the harmonic oscillator potential can be represented by a single particle interaction. There is no way to generate the flat bottom potential with a single particle interaction. A potential that is analogous to the flat bottom potential would be one generated in a cavity of uniform gravitating mass [12], or if you displaced the positive background charges, in the solid state jellium model [27]. Both models represent many-body interactions summed to generate the resulting potential. The advantage in using the flat bottom spherical potential is that it represents a many-body interaction, and not simply a two-body interaction without resorting to perturbation theory. This is the case for both the attractive and repulsive interactions with the particle's medium. What is meant by many-body here is more than one other interacting body [28], three bodies or more.

Summary

Since mass is a key factor in the Schrödinger equation, solutions should exist that illuminate the analytical basis for mass. Encapsu-

lating the many-body interaction of particle to the space it polarizes should define the baseline interaction of the particle and its mass. This physical link between experiment and a quantum description can be exploited. This usually requires either exploring high energy or low energy solutions. Here, we will explore the lowest energy solutions.

1.2 A Short Review

An introductory treatment of quantum mechanics as a subject should be no more difficult than classical mechanics. In the following pages only a small subset of tools from quantum mechanics will actually be used for calculations. Classical mechanics is the basis for quantum mechanics and the clearest reference is one of the early edition of *Classical Mechanics* by H. Goldstein [29]. In classical mechanics, the typical problem is to compute the trajectory of a mass point in some potential with possible secondary constraints. This can be the orbit of a planet or the motion of a vehicle. In quantum mechanics it is traditionally thought that we are limited to measuring the probability density of the object rather than an exact trajectory. In classical mechanics, by using the application of a minimum principles allows a differential equation to be written whose solution will be the trajectory given the starting initial conditions. In quantum mechanics, depending on the particle and the energy there can be two equations that when solved produce something called a wave function. The wave function we will call $u(\mathbf{r}, t)$ whose real absolute value squared is a proportional measure of the local probability density for finding our particle. This is the equivalent of the classical trajectory.

Since the function $u(\mathbf{r}, t)$ can be complex, this density is expressed by the product of u with the complex conjugate of itself, u^*, to produce a real number for the probability density. This is written as $u(\mathbf{r}, t)^* u(\mathbf{r}, t) \geq 0$. You can prove this to yourself, where a and b real numbers, by $(a + ib) \times (a + ib)^* = (a + ib) \times (a - ib) = a^2 + b^2 \geq 0$. When this density function is summed over all space the result should be finite for a single particle and when normalized should equal to 1, indicating the particle exists. This condition is normally written as an integral over all space $\int u^*u d\mathbf{r} \Rightarrow finite$. It is when this result goes to zero or infinity for the particle we know there is a major weakness in the quantum description of the particle that has been proposed. It is a weakness of this type that will be probed for the photon so that a physically acceptable wave function can be set down for that particle. When quantum particles with mass have little energy and are traveling much less than the speed of light, their behavior can be modeled with the Schrödinger equation, which will be used in the next section. It is a misconception that quantum particles are always small and microscopic. If they have very small masses, they can occupy huge volumes and that is something we will consider in detail. If our particle is very energetic and has mass, then a relativistic equation is required, the Dirac equation. The previous reference on classical mechanics also covers special relativity very clearly as that is a fundamental part of classical mechanics, not quantum mechanics. Since we are going to be dealing with low energy free particles mostly in their rest state, we are actually going to have to come up with another equation to understand how things behave in this low energy limit.

Because the solutions of quantum mechanical equations produce a wave function, $u(\mathbf{r}, t)$, we can multiply it by a complex phase factor to give us an extra degree of freedom. This phase factor can be called $e^{i\theta}$. This is allowed because the phase factors complex conjugate is $e^{-i\theta}$ and when it is multiplied by $e^{i\theta}$ it produces $e^0 = 1$. It is as if it were not there. This has been a puzzle to some and is a very important feature of the algebra of wave functions. This extra degree of freedom is something we can use to encode with a particular property as the particle interacts with its environment. This is a feature missing from classical mechanics and we will find essential for defining mass. The actual details of solving the differential equations of quantum mechanics and classical mechanics can be formidable, but the standard solutions are well documented. The good news is that there are only two standard problems: the hydrogen atom and the harmonic oscillator to study. The flat bottom spherical potential problem is a much simpler problem and will be solved in chapter 2. In fact, with the current generation of mathematical, symbolic, differential equation solving programs, it is possible in a few minutes to enter in the equation and plot out families of solutions. This is a good way to experiment with the response of solutions to different parameter inputs.

The simplest quantum problem to consider is the trajectory of a free particle. To get trajectories of quantum particles we need a wave function that is well behaved. This wave function can be turned into a density function that can be measured. If the sum of the density function over all space is finite, the solution describes a quantum particle much as the trajectory of a classical mass point. Quantum mechanics does much more, it describes quantized angular momentum; spin; the creation and annihilation of particles and the phenomenon of extended multi-particle states, which can have very strange behavior from our perspective.

We will treat the vacuum state in a most elementary way, which may offend some, but as a tool it will allow us to use some very elementary techniques of quantum mechanics to illuminate the character of quantum states differentiated by the vacuum state and create the foundation for giving our elementary particles mass. Then, pushing this model, we will then be able to show how mass is an invariant property for any particular elementary particle. The concept of self energy comes out of this development naturally without requiring the use of the Lorentz transformation and with that some more clearly defined notions about both time and measurement appear. There are many books on quantum mechanics and we have singled out two rather old books by Schiff and Dirac [25] [30] as references. Those two books have a very clean mathematical formalism that is minimal and the basics are well explained. The mathematical machinery of quantum field theory is well suited to do detailed perturbation calculations but is not particularly useful in illustrating the origins of mass. The analytical basis for classical mechanics is Newton's three laws, and it appears they make up an even greater part of the foundation of quantum mechanics. Mass unfortunately did not play a major role in the early debates on quantum mechanics as it had been dealt with so recently in the period general relativity was created. The problems facing quantum mechanics that will be dealt with here about mass have changed little from the beginning of special relativity and quantum mechanics.

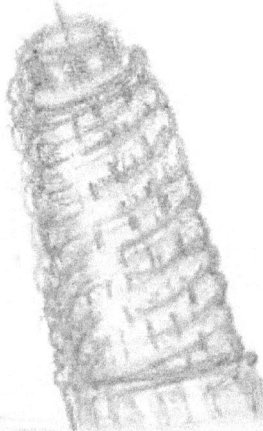

L.M.W.

Chapter 2

Attractive Potential

A particle polarizing space to produce a local attractive interaction can be modeled in three dimensions with a simple potential. It is assumed this will be a weak interaction and it is the solution to this weak polarization interaction that will be sought. The problem of a point mass constrained by a spherical flat bottom potential, depth V and radius a can be solved exactly. There are two regions requiring solutions. The first is in the neighborhood of the origin where the solutions are assumed to remain finite and second is the far region where the potential is assumed to be zero and the solutions decay away toward zero. Our sketch of the potential is shown below and it is not completely filled in yet, but there is enough information to start the problem.

A quantum mechanical solution will provide us with a continuous function that can be complex and is called the time independent part of wave function. In our case let it be called $u(\mathbf{r})$, where the bold faced coordinate $\mathbf{r}$ means the calculation will be done in a three dimensional space. The measurable quantity that the wave function can produce is the probability density distribution of the particle $\rho(\mathbf{r}) = u^*(\mathbf{r})u(\mathbf{r})$. The function $u^*(\mathbf{r})$ is just the complex conjugate of the wave function so that the product and the density is a real function and not complex. The one important thing about the wave function is that it can produce the probability density distribution of the particle over the space it occupies, which is a very statistical measure as opposed to the location of a classical mass point.

A development is found in Schiff, [25] and his notation will be used. Schrödinger's equation for the complete wave function that also includes the time dependent part $\psi(\mathbf{r}, t) = u(\mathbf{r})f(t)$ is:

$$\widehat{H}\psi(r,t) = i\hbar\frac{\partial\psi(r,t)}{\partial t} \tag{2.1}$$

Figure 2.1: **Partial Potential Well of Depth V and Radius** **a**.

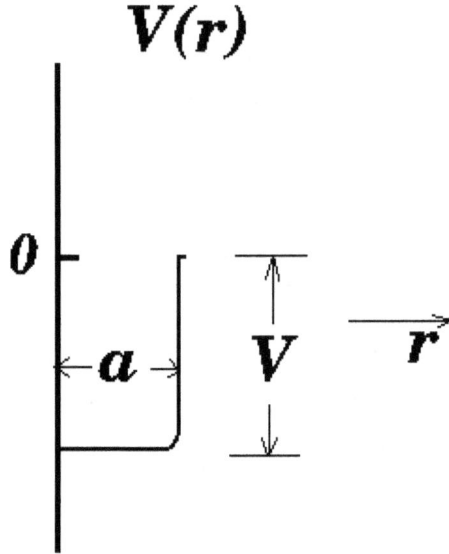

$$\widehat{H} = -\frac{\hbar^2}{2M}\nabla^2 + V(\mathbf{r}) \qquad (2.2)$$

Where $\psi(r,t) = u(r)e^{-\frac{iEt}{\hbar}}$ and in spherical coordinates this results in the three dimensional time independent wave equation:

$$-\frac{\hbar^2}{2M}[\frac{1}{r^2}\frac{\partial}{\partial r}(r^2\frac{\partial}{\partial r}) + \frac{1}{r^2 sin\theta}\frac{\partial}{\partial \theta}(sin\theta\frac{\partial}{\partial \theta}) +$$
$$\frac{1}{r^2 sin^2\theta}\frac{\partial^2}{\partial \phi^2}]u + V(r)u = Eu \qquad (2.3)$$

$$u(r,\theta,\phi) = R(r)Y(\theta,\phi) \qquad (2.4)$$

The angular component of this solutions $Y(\theta,\phi)$ are the very well known spherical harmonic functions that give the symmetries to the atomic orbitals. They are not the principle interest at the moment though they will shape the structure of the higher angular momentum states of particles trapped in the potential well.

Simplifying the constants into two parameters α and β allows the radial part of the wave equation to be written in a more compact form:

$$\alpha = \sqrt{\frac{2M(V - |E|)}{\hbar^2}} \tag{2.5}$$

$$\beta = \sqrt{\frac{2M|E|}{\hbar^2}} \tag{2.6}$$

E is the energy eigenvalue or what we are solving for in addition to wanting to know the particles density distribution. M is the mass of our particle, $\hbar$ is Planck's constant divided by 2π. The final parameter is potential V, the depth of the flat bottom well. These terms are collected together into the two parameter α and β so the calculation can be expressed in a more compact form. Scaling the radial variable with $\xi = \alpha r$ and $\eta = \beta r$, inside and outside of the potential well results in changing only the look of the differential equations and not their character. The order of the derivatives in r of the wave function can be written more compactly as $u' = \frac{du}{dr}$ and $u'' = \frac{d^2u}{dr^2}$. On the interior of the potential well, the radial wave equation becomes:

$$\xi^2 u''(\xi) + 2\xi u'(\xi) + (\xi^2 - l(l+1))u(\xi) = 0 \tag{2.7}$$

This is a very well known equation belonging to the family of equations that generate Bessel functions [31] as solutions. The solutions are made up of the positive and negative half order Bessel functions. The positive set of functions were assumed to be regular at the origin not going to any infinite value and, therefore, are a physically acceptable solution set for the above equation.

$$j_l(\xi) = \sqrt{\frac{\pi}{2\xi}} J_{l+\frac{1}{2}}(\xi) \tag{2.8}$$

The subscripted l represent integers from 0 to ∞ that define the solution set and will then be associated with the particular angular momentum solutions. This integer parameter l will key the radial solution with a particular solution in the angular variables. The first term in this series of Bessel function is:

$$j_0(\xi) = \sqrt{\frac{\pi}{2\xi}} J_{\frac{1}{2}}(\xi) = \sqrt{\frac{\pi}{2}} \frac{sin(\xi)}{\xi} \tag{2.9}$$

subsequent terms are generated from the recursion relation from Watson [32] sec. 3.2:

$$\frac{d}{dz}(z^{-\nu} J_\nu(z)) = -z^{-\nu} J_{\nu+1} \tag{2.10}$$

These solutions are expressible in terms of sines and cosines.

Table 1: Solutions $\xi = \alpha r$ Within the Potential Well

l	u_l for $r < a$
0	$\frac{sin\xi}{\xi}$
1	$\frac{sin\xi}{\xi^2} - \frac{cos\xi}{\xi}$
2	$\frac{3sin\xi}{\xi^3} - \frac{sin\xi}{\xi} - \frac{3cos\xi}{\xi^2}$
3	$\frac{15sin\xi}{\xi^4} - \frac{6sin\xi}{\xi^2} + \frac{cos\xi}{\xi} - \frac{15cos\xi}{\xi^3}$

If $\xi \to 0$ with the approximations for the *sine* and *cosine*, the $l = 3$ solution is not regular at the origin as can be seen by taking ξ to zero . This and a few of the higher order angular momentum Bessel functions were generated from the above recursion relationship and the results were not regular at the origin. It did not appear that the higher angular momentum states would produce a regular value at the origin from the way terms were generated in these Bessel functions, which leaves the three lowest angular momentum states as possible solutions. It is not known whether there are other higher angular momentum solutions that generate a regular solution at the origin. Analytically, higher order der solutions are not expected, which result in just three solutions for a bound state. The solutions correctly satisfied the Bessel equation, which is the most interesting feature of the attractive spherically symmetric flat bottom potential problem. It allows for three different angular momentum states of particles to use this mechanism.

The wave functions are shown in figure 2.2, where the $l = 0$ originates at a value of 1 and the the next two start at zero. All the higher angular momentum $l > 2$ states are undefined at the origin. In addition the density functions are shown in figure 2.3. In a potential free region these wave functions span the entire space and the integral of the density $u^*(\mathbf{r})u(\mathbf{r})$ is not bounded. This implies that these functions must be bounded within a volume to have any physical meaning. This is no different in Cartesian coordinates where the plane wave solutions have the same property and a bounded volume is imposed as an ad hoc secondary condition. Within that volume the particle retains the characteristics of a free particle.

The solutions in the region for $x > a$ are shown in the following table:

Table 2: Solutions $\eta = \beta r$ Within the Potential Barrier

l	u_l for $r > a$
0	$-\dfrac{e^{-\eta}}{\eta}$
1	$i\dfrac{e^{-\eta}}{\eta}\left(1+\dfrac{1}{\eta}\right)$
2	$\dfrac{e^{-\eta}}{\eta}\left(1+\dfrac{3}{\eta}+\dfrac{3}{\eta^2}\right)$
3	$i\dfrac{e^{-\eta}}{\eta}\left(1+\dfrac{6}{\eta}+\dfrac{15}{\eta^2}+\dfrac{15}{\eta^3}\right)$

Figure 2.2: **Wave Function Near Origin for l = 0, 1 and 2.**

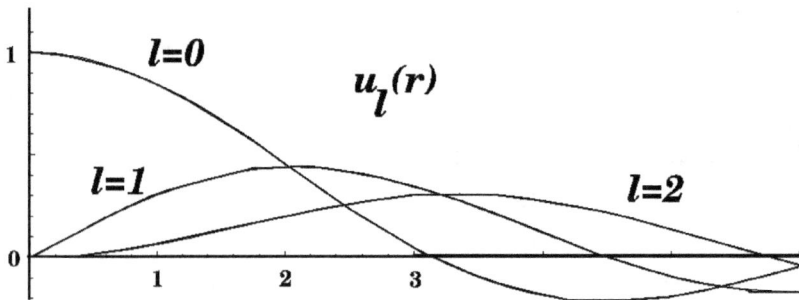

To construct the full wave function the interior wave function must be made continuous with the exterior solutions and when that condition is met the result is an expression for finding the energy eigenvalue of the state.

$$\left(\frac{1}{u}\frac{\partial u}{\partial r}\right)_{r<a} = \left(\frac{1}{u}\frac{\partial u}{\partial r}\right)_{r>a} \; at \; r = a \qquad (2.11)$$

Algebraically there is one feature of this solution that makes the process slightly opaque. The resulting equation is a transcendental equation that is usually solved numerically or graphically by plotting the two functions and picking their intersections. A variation

Figure 2.3: *Probability Density Function Near Origin for*
l = 0, 1 and 2.

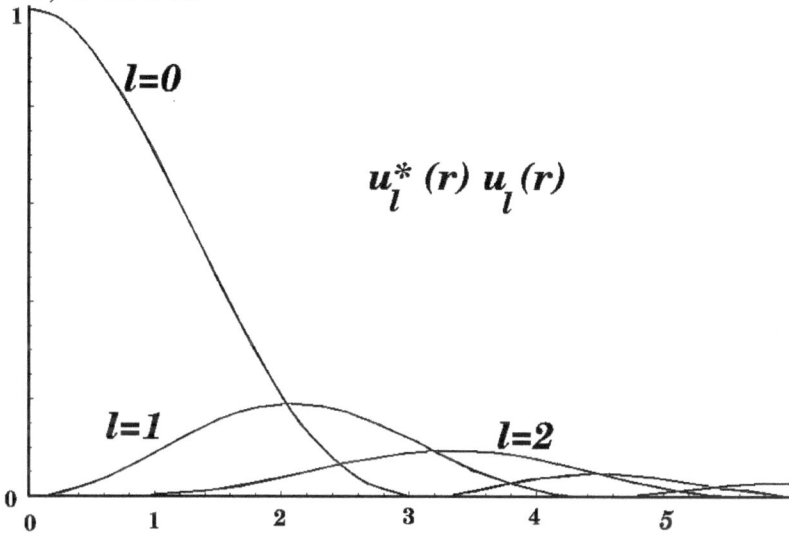

of this method is performed with a mathematics plotting program.
Move all the terms to one side of the equation and divide that into
1 and plot the results. The very sharp peaked functions going to
infinity will mark the solutions. For the $l = 0$ case the above
continuity relation reduces to:

$$tan(\alpha a) = -\frac{\alpha}{\beta} \tag{2.12}$$

From this result, the interest lies in the lowest energy solution for
$l = 0$ which occurs as αa approaches a small number less than
one. The equation is simplified to the form that allows an exact
solution.

$$\beta a = -1 \tag{2.13}$$

Before examining the above relation it is worth while to test $l = 1$ and $l = 2$ continuity relationships. For $l = 1$ the continuity
condition is:

$$\frac{1}{\alpha \tan(\alpha a)} - \frac{1}{\alpha^2 a} = \frac{1}{\beta} + \frac{1}{a\beta^2} \qquad (2.14)$$

in the limit of $\alpha a \ll 1$ the $l = 1$ continuity condition is the same as equation 2.13. Then testing the condition for $l = 2$ where $\eta = \beta a$ and $\xi = \alpha a$ yield the equation:

$$\frac{\tan\xi(4\xi^2 - 9) + \xi(9 - \xi^2)}{\tan\xi(3 - \xi^2) - 3\xi} = -\frac{\eta^3 + 4\eta^2 + 9\eta + 9}{\eta^2 + 3\eta + 3} \qquad (2.15)$$

the condition $\xi = a\alpha \ll 1$ reduces the equation to:

$$\eta^3 + 4\eta^2 + 9\eta + 9 = 3\eta^2 + 9\eta + 9 \qquad (2.16)$$

This relation reduces to:

$$a\beta = -1 \qquad (2.17)$$

Unlike the $l = 0$ solution, which can take on any value for the radius a, the $l = 1$ and 2 require a minimum radius. This is a very curious result as it relates the mass, the well dimension and the eigenvalue. The main implication is that the $l = 0$ solutions can be much more compact than the $l = 1$ or the $l = 2$ solutions and therefore the mass of $m_0 \gg m_1$ or m_2 where the subscript on mass represents the angular momentum.

Writing out expressions 2.17 gives a relation between eigenvalue for these three solutions' result in a simple dependency on mass and the well's radius a. The potential itself is not an explicit part of the relationship.

$$|E| = \frac{\hbar^2}{2a^2 M} \qquad (2.18)$$

The restriction on a is an important condition in order to have the necessary rule for mass. If the space under consideration cannot support a dimension as large as a the particle should be massless. The three lowest angular momentum states have a defined scale threshold that establishes the condition for mass.

There may be an objection that the solution is not valid because mass is defined in a recursive manner with this solution, where mass secures its existence from an assumed condition. The flat bottom spherical potential, as a model for a unit interaction between a particle and the space, represents a scalable mechanism to explore the dependencies for such interactions. Actually, the result defines the minimum volume required for the entity to have the characteristics of mass. Because the solution of this problem relates a scale interaction and mass alone, the analysis is automatically recursive. Mass is now coupled explicitly to a scale parameter. The self consistent solution of the spherical flat bottom potential shows that mass is not an intrinsic property; rather, it is an acquired property requiring a volume of space with an attractive interaction resulting in three stable angular momentum solutions.

So for $l = 0$ with an attractive interaction, there are no further restrictions, on the interaction to have a physically acceptable wave function. As the angular momentum is increased to $l = 1$ and 2, a restriction comes into play where the potential well must exceed a minimum diameter in order to have a well behaved description. The immediate implication of the increased volume required for $l > 0$ states is simply that the $l = 0$ state has no minimum restriction on the well radius and therefore can be a very heavy boson state. Whereas for $l > 0$, there is a restriction on the minimum size and it would be a much less massive entity. If there was a particle that could exploit all three states, there could be one heavy version and two light versions.

Assuming the attractive potential volume shrinks toward zero, the conditions for at least one bound state, $l = 0$ can be maintained. The condition gives an unambiguous definition of a stable particle without resorting to mechanisms like defining a wave packet to truncate the wave function at large distances. To have symmetries that are non-spherical, $l = 1$ & 2, the volume required to define a massive boson has a minimum volume limitation greater than that for $l = 0$. The basis for the wave-particle duality properties that are evident in experimental measurements may result from a hierarchy of interactions, one which generates mass over a volume and then secondary interactions with external potentials and structures.

Summary

We discovered a simple new solution to an old quantum mechanical problem. The limiting weak attractive interaction has only three possible allowed bound states with angular momentum 0, 1 and 2. The condition on the eigenvalue is different for these states compared to the normally described bound states because the potential well depth is not part of the eigenvalue condition. The relationship is between the eigenvalue, mass and the well's radius. The eigenvalue relation is close to a definition of what is required for mass to arise from a local attractive polarization. This type of interaction describes the properties required for a boson that can collect multiple particles within the same state. However, below the minimum scale of the potential well the particle will appear to be massless if measuring in an experiment below that scale. This set of solutions points to an incomplete description as provided by the Schrödinger equation and has to be viewed as a defect in the description of mass that can be corrected. It, however, links the vacuum interaction to a particle in a simple way defining mass as a condition for a solution to exist.

2.1   The Mass-Scale Relation

Our initial assumption states that a particle in space polarizes the vacuum or some region in the solid state and produces an attractive local potential. This act strains the simple model we are using, because it is now going to be used to describe an unstable response. The mass defined from the limiting condition, equation 2.22, in the continuity relationship can be simplified. The small energy $|E|$ can be defined in a period τ to that precision by the energy uncertainty relationship.

$$|E|\tau \geq \frac{\hbar}{2} \tag{2.19}$$

Eliminating the energy eigenvalue yields an equation that defines the upper bound in mass.

$$M \leq \frac{\hbar\tau}{a^2} \tag{2.20}$$

The measurement period can be no less than the time it takes for light to cross the potential well.

$$\tau = \frac{a}{c} \tag{2.21}$$

This leaves an upper bound on mass that is only dependent on the dimension of the well.

$$M \leq \frac{\hbar}{ac} \tag{2.22}$$

The upper bound value of the mass becomes

$$M = \frac{\hbar}{ac} \tag{2.23}$$

Mass is proportional to the curvature of the site and mass is a property of the site as a whole defined only for scales of a or greater. Since mass is an upper bound and the solution is a conditional statement on mass and the eigenvalue, then measurements done on a smaller scale will find that the particle is of a reduced mass or effectively massless if the scale is many orders of magnitude less than that of the potential well's radius.

This last condition, along with the allowed angular momentum states mirror the properties of the photon for an $l = 1$ particle. The pAxion in some measurements appears to be an example of $l = 0$ form of the interaction. Before looking at these cases in detail and searching for a potential that can support such behavior, it is necessary to flip the potential into the positive domain to look at what eigenfunctions are allowed for a repulsive potential. This representation also occurs as the expression for the Compton wave

length and differs from the mass in a classical field because now
an angular momentum value has been joined to its definition.

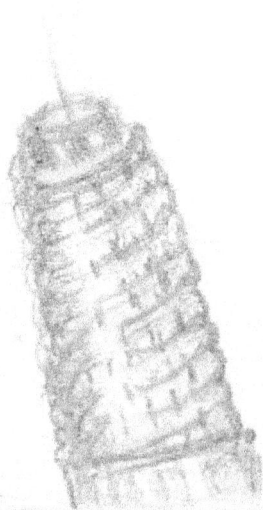

L.M.W.

Chapter 3

Repulsive Potential

Flip the attractive negative potential positive to create a repulsive potential, but what can be done with it is much more limited. The repulsive potential is required at the origin to model a repulsive polarization and then the integrated density of the solution must be finite to define a valid particle. This second condition is a restriction on what kind of potentials are allowed. For the start of the problem, it will begin with a partial sketch of the repulsive potential, but for there to be a finite value for the total particle density, the potential will have to rise away from the origin producing a bound state. That raises the question is the process being well described. Starting with this simple model, each revision will be taken as far as it seems useful.

Figure 3.1: *Core of the Repulsive Potential*

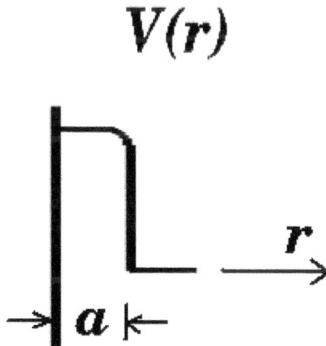

There are no assumptions about the repulsive potential's magnitude, but we have the ability to vary both its strength and its radius. There is no reason to assume the interaction is weak as was done for the attractive potential. Since this is a potential that will be generated in response to the particles presence, it is assumed initially to generate a random response from our particle.

This series of random response is expected to destroy and angular symmetry for the wave function of our particle. So the search will be for a wave function with $l = 0$. There is nothing in that space outside of the potential to select a particular symmetry that breaks the spherical symmetry of the initial conditions. The interior solution is a decaying exponential and a decaying sinusoidal on the exterior. As a for the exterior solution shrinks, the solution should not grow unbounded as the interaction is repulsive. This is taken care of by the $l = 0$ solution from Table 1.

$$u(r) = \frac{A}{r} sin(\beta r) \qquad (3.1)$$

Within the repulsive potential the wave function must remain finite and that results in the expression, for $r < a$:

$$u(r) = B\frac{sinh(\gamma r)}{r} \qquad (3.2)$$

The scale parameters are:

$$\gamma = \sqrt{\frac{2M(V - E)}{\hbar^2}} \text{ and } \beta = \sqrt{\frac{2ME}{\hbar^2}} \qquad (3.3)$$

The continuity relation result in the expression:

$$\beta cot(\beta a) - \frac{1}{a} = \gamma\frac{e^{\gamma a} + e^{-\gamma a}}{e^{\gamma a} - e^{-\gamma a}} - \frac{1}{a} \qquad (3.4)$$

which simplifies to the relationship

$$\frac{tan(\beta a)}{\beta} = \frac{tanh(\gamma a)}{\gamma} \qquad (3.5)$$

If the condition of forcing the arguments to be small relative to 1 can be made, then the condition results in the simple expression which is true:

$$a = a \qquad (3.6)$$

This almost appears as an ambiguous solution, but it contains a good deal of information. Before studying the result a check of the continuity condition for the two other possibly allowed state of $l = 1$ and $l = 2$ will be done. The wave functions for the region of $r < a$ are listed in Appendix A. For the case of $l = 1$ there is no solution in the limit as r goes to zero because the result is $a = -a$. However, for the solution for $l = 2$ the result is the same as for $l = 0$. The higher angular momentum components in the case of the repulsive potential as opposed to the attractive potential are probably not realistic solutions because they do not form a stable state as does the attractive case. The continuous forming and scattering is a random process and the best that can be captured from this process is a randomized state with no angular symmetry.

The solution appears to be independent of the wave function's phase in each region. A degree of freedom that allows the parameters that make up the phase so it scales so the net phase remains constant. Since the phase is small, it will be designated δ for the γ term.

$$\delta = \gamma(a)a = a\sqrt{\frac{2M(V - E)}{\hbar^2}} \qquad (3.7)$$

$$\beta(a)a = a\sqrt{\frac{2ME}{\hbar^2}} \qquad (3.8)$$

This solution can appear to be a paradox if examined in the frame of reference of the particle from which it is trying to escape. If it is instantaneous or a single event, it is not very interesting, but the sequential reforming of the repulsive potential requires the dynamic equations to include these delayed responses. This recursive break down of repulsion is much like an image from the fable of the sorcerer's apprentice, one has two creatures a programmed burrowing rodent, an energetic mole, who lives underground and burrows up under his surface living counter part displacing him randomly and following him underground when the fellow on the surface jumps to a new site with the process repeating with the burrowing rodent knocking his friend off balance. The features of this process are again the strength and physical range of the interaction.

As $E \to 0$ and the only interesting expression is

$$\delta = a\sqrt{\frac{2MV}{\hbar^2}} \qquad (3.9)$$

The question becomes whether the phase δ is a constant, independent of the scaling of a. If the phase is constant, then the interaction scales in such a way its energy is also a constant and independent of a. If this is a charge caused polarization space, then the potential V will scale inversely with a. This relation is analogous to the relation generated for the attractive potential. If this quantity is a phase, it has to be a constant . Then the product $MV \sim \frac{1}{a^2}$ for a charged interaction with V is proportional to $\frac{1}{a}$ which shows mass the same inverse dependence in a as a boson.

The mass can be solved for and results in:

$$M = \frac{\delta^2 \hbar^2}{2a^2 V} \qquad (3.10)$$

The unknown quantity in this expression is δ. To reduce this derivation to a form for a charged particle, the bare electrostatic potential is used and the expression for mass becomes:

$$M = \frac{\delta^2 \hbar^2 \epsilon_o}{2ae^2} \qquad (3.11)$$

Here mass is again inversely proportional to the scale parameter, a. For the electrostatic potential the value of the phase is approximately $\delta \sim 10^{-7}$ a small number.

The particle's interaction with the vacuum state initiates a process in a repulsive manner generating a dynamically stable chain of events. There is an implied delay in this process that allows it to take place and that is captured in the phase δ. Unlike the solution for the attractive potential, which must be weak, this can be a strong interaction to the repulsion from the perturbation of the space. This dynamic perturbation process to an observer over time appears as a single continuous interaction. The potential interaction magnitude will set an upper limit on the mass of a stable particle. Unlike the case of the boson where a fixed value of a was generated, this interaction can occur at all scales. A microscopic analysis of this process is necessary as this derivation was just to show what essential conditions are necessary for the mass to be derived from a continuous repulsive interaction. In both these calculations, the spherical flat bottom potential was used as a simple unit potential that could be considered deformable in order to explore the conditions necessary to establish the characteristic of mass for a boson and fermion.

Summary

The potential interactions are much more complex for a local repulsive interaction and originate in the strong role the many particle interaction can play with a charged particle. There are also implications that the long range normal potential behavior for particle with mass will be constrained to specific types. Mass by this mechanism of repulsion can be defined at all scales which differs from the attractive potential. The source of these potentials are due to mechanisms specific to the particle involved. An analytical picture of the repulsive interaction requires a statistical description.

3.1   Symmetry and Spin

The topic of symmetry in quantum mechanics is well developed and useful in solving a number of problem in quantum mechanics [33]. Our approach in the use of symmetry is rather practical and cautious as we are initially not going to assume a complete knowledge of the forces that produce measurable symmetries. This caution is based on the understanding that it is not always possible to clearly separate the sources of multiple symmetries found in measurements until the problem is completely solved. For example if in a particular system angular momentum is conserved it can be said that requires the space to be isotropic with respect to rotation. That is true for that particular system where angular momentum is conserved. It would be a bold assumption to extend that property to all particles no matter the type in the same space. Even for the particles where angular momentum is conserved there might be a specific identifiable reason why the space is isotropic for that system rather than just relying on an assumption of isotropy for any type of particle in that space. Some current analysis in physics gives weight to symmetry arguments assuming that they will transcend effects which may not be well defined or understood. This type of argument is generating new conservation conditions. In the solid state none of these assumptions are very useful unless they are checked for specific systems. For example chiral effects are usually the result of the action of two different force types. The presence of chiral structures is an indicator of an order application of two different forces, possibly a central force and axial force operating on a structure. One classical result is the Coriolis effect whereas a quantum effect would be found in the beta decay in Co^{58} or Co^{60}. The product of the action of the forces which generate the chiral characteristic gives some history of the process on a non simple structure. How general the process is can only be determined by finding many examples with some understanding of the details of the process. However, in non isotropic spaces like solids or in vacuum where the details of the interactions may not be completely understood it is useful to view symmetries as only a product of the allowed interactions. There is a theorem in thermodynamics called Nernst theorem that deals with the limited role symmetry plays in the energies allowed by a system as the system's energy is reduced [34]. Since we are considering systems at very low energies approaching a rest energy, these thermodynamic considerations may not be out of place when considering the forces that will shape the low energy states; therefore, we will assume that physically discernible symmetries are the result of action by forces on structures, and not an a priori constraint on actions themselves.

The change of sign in the local potential interaction of a particle with space from positive to negative affects the problem's fundamental symmetry. The attractive interaction always remains centered on the particle maintaining a spherical symmetry of the wave function even in the presence of other fields. The repulsive symmetry is an unstable symmetry and can be perturbed to take various trajectories. The trajectory, defined by a vector **r**, automatically defines the untaken trajectory in the direction **r̄**. The repulsive interaction, when in the presence of an additional field, will lead to a bifurcation of the initial state and a reduction in symmetry. This is a feature that would be more characteristic of fermions than bosons.

Figure 3.2: **Symmetry of Force Interactions For Attraction and Repulsion**

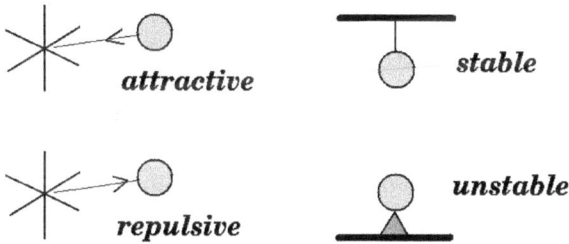

The interesting solution for the repulsive potential was found for the $l = 0$ angular momentum, which allows $a \to 0$. The other major difference is that mass is defined all the way to the limiting minimum dimension with no cut off value. This is very different than the case for the attractive potential well where there was a non zero dimension to define the wave function.

Table 3: Mass Characteristics for Fermions and Bosons

Potential	Conditions	L	Comments
Attractive	$M = \frac{\hbar}{ac}$ for $r > a$	0,1,2	bosons
Repulsive	$M > 0$ for $r > 0$ $V \sim \frac{1}{r}$ then $M = \frac{k}{a}$	0	fermions

The second feature that was not discussed in detail for the repulsive potential is that the total density of the wave function must remain finite which requires the potential's magnitude away from the origin returning to zero. A crude schematic of such a scaling potential would look like the following figure.

Figure 3.3: *Next Approximation for the Repulsive Potential*

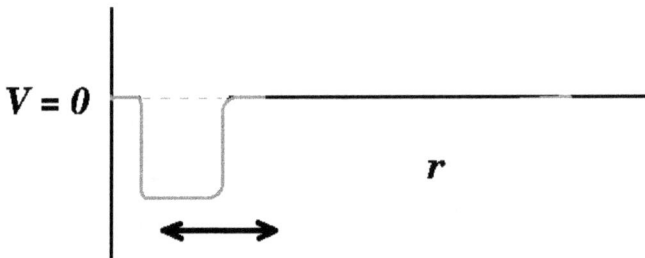

This in three dimensions generates a spherical shell that is a potential minimum to confine the motion of the particle in response to any external potentials. The spherical degrees of freedom allows the orientation of any angular momentum captured in the motion to respond freely to an external field. The detailed shape of the potential will depend on the interaction, but the basic form allowing an internal angular momentum states will not change.

Summary

Finding that the origin of a boson or a fermion may be no more complex than the sign of the polarization of the space the particle occupies was important. The repulsive potential, polarization interaction appears to require a charged particle to generate an effective spherical shell potential trough around the original center of repulsion. Because of the unstable nature of the repulsive potential the symmetry of the response to external potentials will give rise to a splitting of states with two possible trajectories that would be characteristic of fermions. Only when the detailed behavior is shown compatible to the potentials generated can this point be proven.

Chapter 4

The Differential Equations of Quantum Mechanics

"It should be emphasized that we have not given a rigorous deriva-
tion of the Schrödinger time-dependent equation from classical me-
chanics. No such derivation exists, any more than it does for the
time-free equation previously considered" Stanley Raimes page 312
"The Wave Mechanics of Electrons in Metals" [35]

The above quote highlights the basic problem with mass. Both the Schrödinger and Dirac equations have mass as a key element and particular solutions of the Schrödinger equation relate mass to a spatial scale. The derivation of the Schrödinger equation should also hold the basis for the the description of mass. The justification for this point of view is simply the overwhelming utility of the equation itself. Therefore, a derivation that is consistent should also contain significant information about the origins of mass.

To this point we have modeled the interaction of a particle to a vacuum or medium using the concept of a time averaged potential with no microscopic description. This effort yields some simple symmetries in these minimum potential solutions with a very well defined finite set of solutions. The description was done with a particularly simple potential that was allowed to scale in its parameters so that a self consistent definition of mass would emerge. To extend this model to a more detailed level of interaction requires a microscopic description of the polarization of the space. This reduction in analysis scale immediately brings in a degree of randomness that is essential in sharpening the concepts about the origins of mass. The interaction with vacuum for a boson, results in two effects to apply to the potential model previously described. First, there must be an attractive interaction and second, these interactions should produce a type of quantum Brownian motion. These references begin with Einstein's view of Brownian motion

[36] and how it developed in a review by Kac [37], they were then incorporation into quantum mechanics for the harmonic oscillator by Schwinger [38] and for the $\frac{1}{r}$ potential by Gutzwiller [39]. The one comment that can be made on the quantum nature of Brownian motion to be discussed here is that it appears as a very weak perturbation on the scale of the observer and where the dynamic equations are solved. This weakly observable quantum Brownian motion uses the mean position as the frame of reference. The classical Brownian motion generates the diffusion equation, and similarly the quantum mechanical version is expected to generate a differential equation that will be related to the Schrödinger and Dirac equations. Most of the previous studies of quantum Brownian motion have dealt with the harmonic oscillator properties or orbit trajectories of the $\frac{1}{r}$ potential following the development of the harmonic oscillator and the hydrogen atom solutions. Here we try to examine the trajectory of a free particle, an in particular its rest state.

The Schrödinger equation is typically motivated by a correspondence between a classical Hamiltonian and a quantum mechanical version that satisfies the algebra of the allowed solutions for a single particle problem. The simplest problem that is considered is a single propagating particle stripped of any applied external potentials. Stated another way, any spatially dependent potential components average in such a way that they do not appear explicitly in the final relationship which is the best the single particle approximation can produce. This results in a model for the particle's dispersion relation, $E(\mathbf{p})$ or $\mathbf{p}(E)$, relating momentum to energy. This is a reduction in the number of variables to only the momentum with no spatially dependent potential components affecting properties, or more simply Newton's laws applied to a quantum particle. The remaining problem is that mass is added in an ad hoc fashion in this development of the Schrödinger equation. Initially, to avoid an arbitrary addition of mass to the model we will examine the case of a massless boson. On a small scale for the photon we have Maxwell's field theory where the aim is to develop the additions to the dispersion curve from the interaction between the photon and the vacuum. For the repulsive potential, the electron is a model system. If the particle-vacuum interaction can capture features of the random processes, there may be a formal way to derive a descriptive equation directly starting from a massless particle's dispersion relationship.

Summary

The conflict in using the Schrödinger equation to provide a self consistent definition of mass comes from the confusion about defining a quantum description that is anchored to the frame of the particle. Whether this is possible or not is the subject of what follows. There are two approaches to this problem, one is to add up all the raw interactions or alternatively to encapsulate the raw interaction into a potential. The latter approach is favored here because it may not be possible to identify all the individual interactions. [40].

4.1 Boson Phase Delays

The photon will interact with the vacuum or another boson within the solid state interacts with medium in which it propagates and these interactions form the baseline properties of the measured properties of the particle. The lowest order interaction of a boson is represented in figure 4.1.

This interaction for a photon's electric field with electron-positron pair is highly improbable because it produces a final phase shift in the photon of $-\pi$. The response of the electron-positron pair is to re-radiate with the opposite phase and with a complete loss of momentum conservation. Since we don't see a reflection from a passing visible light in vacuum, this is evidence that the process violates the conservation of linear momentum.

An interaction in order to produce the effect of an attractive potential has to leave the phase of the photon with a negative phase shift relative to a non-interacting photon. The diagram that is required for producing a negative phase shift, and hence

Figure 4.1: ***Freely propagating photon and a transient vacuum effect of a photon generating an electron-positron pair.***

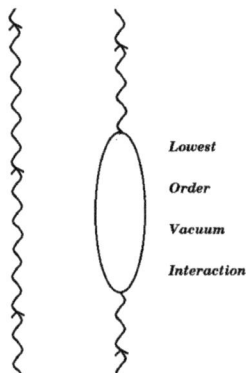

Lowest

Order

Vacuum

Interaction

an attractive potential, requires that the accumulation of phase shifts are reduced to small values in the process while not violating conservation of momentum and energy. The basic diagram is that of parametric down conversion, an event that occurs in nonlinear media that is only usually visible with high amplitude photon fields or in non-linear materials that can allow the process to go to completion as the material participates allowing energy and momentum to be conserved. The base diagram is shown in figure 4.2.

In a dilute medium like vacuum this is a highly improbable event because the final two photon states are not favored because those final states were not previously occupied and there is still a phase shift of $-\pi$ in the final state that violates overall momentum conservation. The interesting diagrams for the introduction of quantum Brownian motion for a boson, and a photon in particular are found in figure 4.3 and 4.4. These processes complete a conserved process of returning a photon to its initial energy state, with a slight phase delay. This phase delay is equivalent to a random displacement $\Delta\theta \sim -\Delta x \mathbf{p}$.

These events in vacuum are not expected to have a high probability, but they result in random phase delays, ($\Delta\theta$ for a photon)

Figure 4.2: **Basic Diagram for Parametric Down Conversion**

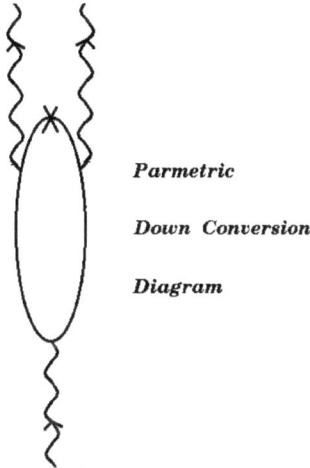

Parmetric

Down Conversion

Diagram

Figure 4.3: **Basic Diagram for Photon Phase Delay**

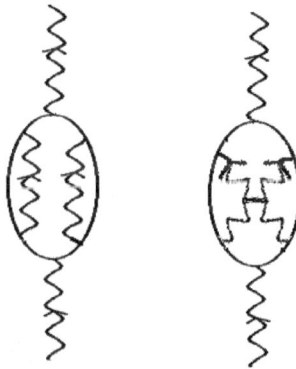

$$w \rightarrow w_1 + w_2 \rightarrow w$$

Phase Delayed Boson Diagrams

because the rate of phase accumulation in the split state is on average one half that of the original state and their frequency is one half that of the original state. This random phase delay can also be described as the effects of a local interaction with an effective attractive potential.

$$\Delta\theta \sim \langle \Delta t \rangle (\frac{\omega_1 + \omega_2}{2} - \omega) = -\langle \Delta t \rangle \frac{\omega}{2} \qquad (4.1)$$

This net delay acquired through the complete process is a result of the bosons interaction with a transient plasma and its coupling with any matter, the most likely source of the bosons attractive potential to the vacuum state. These events for the photon or boson field in vacuum require long periods of time as compared to activity in highly non-linear media which requires short periods because of the stronger coupling to the medium in the solid state. The quantity a that was previously computed represents the minimum mean linear size of the volume required for this polarization interaction. The photon phase displacement is introducing a random displacement in the field of the photon. The result will be a phase shift that will satisfy the requirements of energy and momentum conservation along with producing an effective attractive weak potential. On a size scales larger than a one is forced to a probabilistic description of the photons behavior resulting in random phase additions yielding a mass to describe the photon's interactions in space.

The last two processes only changed the phase of the boson randomly. Also, from that state it is possible to radiate particularly if there are states capable of accepting that radiation. The diagram can be redrawn including two extra photons that are generated. This process in vacuum is more improbable, but in a dense high-temperature plasma, if the lower energy states are populated, this process may be channel for high energy gamma rays to lose energy.

In the limit as the frequency approaches zero for the photon pair radiated, the above argument breaks down unless there is a finite self energy floor beneath which the frequency cannot be reduced. If the previous conserved phase delays generate mass then that defines the minimum self energy of the photon and the minimum frequency. If this is the case the upper bound of the radiation emission rate can be estimated from the transition probability for

Figure 4.4: *Photon Phase Delay and Radiating*

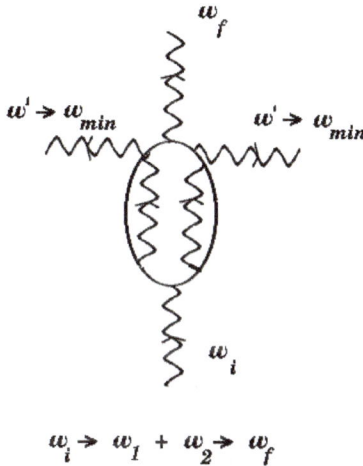

$$w_i \rightarrow w_1 + w_2 \rightarrow w_f$$

Phase Delayed Boson Diagrams
Conserving energy
and momentum

spontaneous emission to $\hbar\omega = E_k - E_n$. Assuming E_k is very high energy compared to the minimum energy photon, then ω is a very small number and the transition rate per unit time is simply [25] given by the following expression. Where $(r)_{kn}$ is the dipole matrix element between the k and the n state.

$$\frac{4e^2k^2\omega}{3\hbar c}|(r)_{kn}|^2 \tag{4.2}$$

If the matrix element to an allowed state can be found to be non zero, then there would be an efficient mechanism to populate the very low energy photon states by very high energy photons. This diagram could become very active at the center of a dense hot star with temperatures at $10^9 K$, where there would be a dense electron-positron-gamma ray population available to drive the process. In addition, if there was a significant population of minimum energy photons, there could be a significant transfer of energy into these low energy photon states. This diagram can also be active when the path of light is redirected by an angular momentum transfer via a minimum energy photon that results from a strong gravitational potential altering the path of the original photon. This is a microscopic description of light following the gravitationally curved space of general relativity.

There has been no mention made of the requirements of the state of the entire system for these processes to take place efficiently. In order for there to be an exchange of momentum the vacuum must not be empty but have a finite density of matter. The rate of interaction of a photon with this matter will depend upon both the density and the temperature of the matter and the background radiation field. The limiting value for the existence of mass is found in equation 2.22 as $M \leq \frac{\hbar}{ac}$, as an upper bound on mass, because the solution was for the minimum volume required to define a boson with mass. The relations worked out from the vacuum phase delay will result in mass being defined as a lower bound.

$$\Delta p \Delta x = \frac{mc^2 a}{2c} \geq \frac{\hbar}{2} \qquad (4.3)$$

yielding:

$$m \geq \frac{\hbar}{ac} \qquad (4.4)$$

These relations result in a definition of mass identical to $m = \frac{\hbar}{ac}$ for the mass of the photon.

In the case of the pAxion, a linearly polarized magnetic excitation, its interaction with the medium is magnetic. Unlike the interaction of a photon where the electron-positron pair lowest order response does not satisfy momentum conservation. The pAxion interacting via a long range current rather than spin wave excitation which are massive excitations is more probable. Like the photon, the pAxion is a light particle interacting and can only weakly interact with a heavy particle. The major property change that occurs for the pAxion-pAxion interactions above the Curie point is that momentum transfers no longer couple directly to the lattice via the lattice's lost magnetization. This switching off of the bulk magnetization above the Curie point produces two results. First the phase delay per unit distance propagated is reduced and second the ability to form a Bose-Einstein condensation is lost.

Summary

The photon's phase can be altered by interactions in space. Only if the phase can be retarded in a natural manner can the photon appear to have mass. On the laboratory scale we know this is not the case, however, on larger scales it is thought that it may be possible that the photon will acquire a mass [15]. The sketch of the mechanism introduced is something common from quantum optics in nonlinear crystals. If the rate of these interactions are sufficient, they will be detected in a photon as mass. Such experiments will be limited to astrophysical scales for the photon where the phase can be altered and also energy lost to local potentials. The phase is important because it is a variable that is like an appendix to the quantum description of particles and fields. If it is partially a random variable with retarded characteristics, it then becomes the mathematical description of the mechanism by which mass can be acquired.

4.2   Fermion Random Walk

The problem now jumps from entities with a mass much less than 10^{-54} kg an estimated upper bound for a photon mass and 10^{-39}kg for the pAxion mass to electrons, the simplest elementary fermion along with its anti-particle, at the relatively heavy mass of 10^{-30}kg. The lowest order interaction of a charged fermion that represents a version of quantum mechanical Brownian motion is shown in the figure 4.5.

It is the dynamics of the process leading up to the annihilation that determines the effective potential the electron experiences.

Figure 4.5: *Fermion Vacuum Interaction*

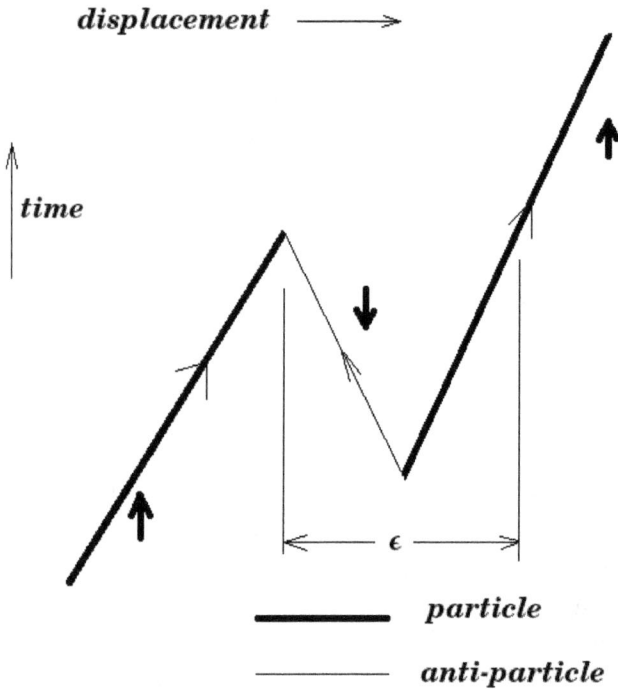

The initial polarization response to an energy matched electron-positron pair represents the interaction that will determine the properties of the electron's random walk. A long time scale version of this problem would follow the development for plasma oscillations, but since the process between the electron and the vacuum excitations is a terminated interaction only the initial time development of the polarization properties are important.

The first order potentials for its vacuum interaction of our initially free electron e_1 are summed in V_1, where the arrows represent the spin states.

$$V_1 = V_{e_1\uparrow - p\downarrow} + V_{e_1\uparrow - e_2\uparrow} + \\ V_{e_1\uparrow - p\uparrow} + V_{e_1\uparrow - e_2\downarrow} \tag{4.5}$$

Since the initial electron can only annihilate with the anti-parallel spin positron the reduced potential V_1^R is made up of two terms.

$$V_1^R = V_{e_1\uparrow - p\downarrow} + V_{e_1\uparrow - e_2\uparrow} \qquad (4.6)$$

What is important for $V_1^R(t)$ is its time dependence as the polarization changes the contributions of the three terms in the total potential. The polarization among all three components will drive them toward a different total potential. It is assumed that initially for any type of meaningful reaction rate, the three wave functions are sharing the same volume. In the polarization process the mean values of the potentials should obey the inequality: $|V_{e_1\uparrow - p\downarrow}| > |V_{e_1\uparrow - e_2\uparrow}|$. This inequality is the source of two features of the potential. The core appears repulsive and a spherical shell of potential forms that supplies a bound state as the potential vanishes for large radius and a bounded density integral for the wave function. A sketch of the potential is shown below:

Figure 4.6: **Improved Fermion Potential Interaction in Vacuum**

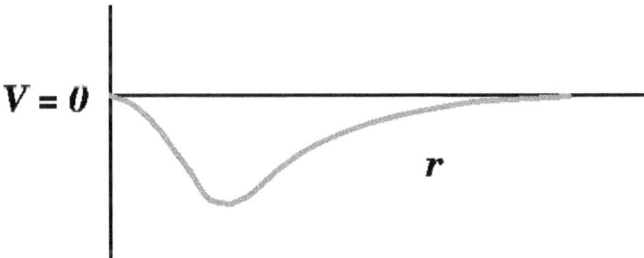

In order to generate this potential accurately, the ground state wave function of a particle with mass at rest is required for the three interacting particles. Using the arguments developed here, a quantum wave equation will be derived in the next section that will allow this relaxation potential to be computed.

A building repulsive potential is encountered by the original electron that can be annihilated. This can only occur when the initial electron can undergo a singlet annihilation with the positron.

The net effect of annihilation is a displacement. So for this particular fermion, the conditions for a repulsive interaction drives
a polarization ending in a possible annihilation event that locks
in the effective displacement and then the process repeats. The
mechanism generating the spin state for the initial electron also
operates for the virtual electron-positron pair.

Unlike the case of the photon, the random properties are introduced directly in this process by the displaced annihilation process. It is not the original particle that is displaced, but it is the
replacement by a displaced virtual electron with the elimination
of the original electron by its antiparticle. This appears to be the
appropriate way to determine the effective displacement of the interaction ϵ. The weaker this interaction the greater the range and
the lower the mass. Both these interactions depend on the time development to create the potential. Because of the strong potentials
of the components involved and the ability through polarization
to form a deep potential well, this interaction with the vacuum
state can be strongly driven by the process itself to establish not
only the mass of the particle which can be significant but its spin
and the associated effective exchange interaction properties in the
same process.

Then the question remains: what is the source of the exchange
effect because it cannot be added a priori. The exchange process
appears to be due to the dynamics of both the polarization of
possible vacuum electron-positron pairs and their allowed interactions with two initial electrons. In the case of two anti-parallel
electrons, there are two independent exchange annihilations possible among the set increasing the total rate of the process which
allows the process to span the volume of interaction. With two
parallel spin states if there was an exchange annihilation, then
the second annihilation event would be conditional on the first,
which makes the process less likely in the case of the parallel spins
increasing the interaction interval and allowing the ensemble to
separate via electrostatic repulsion. In effect, the correlation in
the anti-parallel spin case is a product of the allowed possible
joint exchange annihilations yielding an effective displacement of

the original electrons, and the result is a dynamic correlation of motion. This argument ignores any magnetic field contributions to the process that will also drive the effective correlation dynamics of the process.

The same kind of repulsive potential, driven relationship cannot be constructed for a charge free fermion to generate the conditions for a spin state. This requires another operating repulsive potential to first generate the intrinsic spin of the particle. This description does not generate a fermion for an elementary neutral particles. The most common neutral spin one half particle is the neutron and it is a composite particle that is not stable as a free particle. The second most common in this line is the neutrino which is complex with three states that are available and this may imply it is also a composite particle [41]. There are other unstable neutral fermions which have complex behaviors that may indicate they must be treated as composite particles. The potential arguments we have made do not directly apply to composite particles until the mechanics are described at the level of the individual components.

Summary

To deal with the microscopic repulsive interaction of a particle and its space, we have had to assume a charged species and the direct interaction with the electron-positron sea of virtual particles. The strength of this interaction which will be partially field dependent will determine the mass of the particle. We are moving up the mass scale compared to the photon and the pAxion. The interesting deduction is that neutral particles classified as fermions are more likely to be composite particles. This was a result of the inability to identify a pure repulsive force in additions to a charge based force.

4.3   Quantum Brownian Motion

The professor who started with quantum mechanics for freshman, or anyone motivating the derivation of the wave equation from the energy relationship of special relativity alone, is taking a risk of missing something very basic. Typically it is often taught that Newton's laws describe the behavior of the "large and slow". In particular, it is easy to ignore the scope of Newton's first law: A particle at rest will stay at rest and a particle in uniform motion with no forces acting on it will remain in uniform motion. Quantum mechanics currently is not considering the existence of the rest state. Though the energy relationships that motivate the Lagrangian and Hamiltonian representations appear analytically functional, the basic concepts of particle dynamics as embodied in Newton's laws must be extended to quantum particles because there is no small scale limit for Newton's first law. This has been a fundamental problem in quantum mechanics from its beginning, though we are not aware that it is actually stated in these terms. Its importance becomes apparent when we take an analytical look at what is meant by quantum Brownian motion. Classical Brownian motion was the observation of tiny pollen grains trapped by the surface tension of water, which moved visibly under high magnification because of statistically random scattering event by some highly energetic molecules striking the grains [36]. The solution to this early puzzle shows that water is actually made of molecules and not a linear isotropic and homogeneous and static medium. It appears that arguments concerning classical Brownian motion have to be extended to quantum mechanics not for showing the existence of molecules but to generate the mechanism that yields mass. Examples have been made of only two particles showing how a description of random events have to be involved in the description of photons and electrons as examples of how bosons and fermions respectively can acquire mass. Introducing these elements of a phase delay for a photon in a volume and random displacement of the electron requires an analytical description that includes these effects. There is an engineering analog that treats

these kinds of problems in time. This description can be altered
to treat this problem in space. If a particle-space interaction is dependent on a time delay [42] it also can be represented as a spatial
displacement, responding to the gross behavior of the particle to
represent either a phase delay or an interaction with a vacuum particle and anti-particle pair. This model is forcing a single-particle
model on an inherently many-particle problem and encapsulat-
ing the many-particle contribution in the delayed response of the
space.

The development of the relativistic wave equation for the electrons typically start with the relativistic definition of the energy
momentum relationship.

$$c^2 p^2 + m^2 c^4 = E^2 \qquad (4.7)$$

This equation can be simplified for massless particles and could
provide a starting point. However, the Klein-Gordon equation
that can be derived from this expression has problem solutions
with time dependent probability densities that are not valid particle states [43]. Also, our principle interest lies with low energy
solutions in the frame of reference of the particle. In particular it
is the rest state solutions that are the goal in this section which
to the first approximation is not a relativistic problem. The rest
state problem cannot be done explicitly for either the Schrödinger
equation or the Dirac equation. So we will start with the general
definition of Newton's third law, the proportionality between the
rate of change of momentum and force coupled with a use of his
first law to develop the equations required.

$$\frac{d\mathbf{p}}{dt} = F = Ma \qquad (4.8)$$

We are going to apply Newton's third law first to a particle free
of external forces. It would be free of external forces if it is in a
constant potential region of space.

$$\frac{d\mathbf{p}}{dt} = 0 = \frac{dx}{dt}\frac{d\mathbf{p}}{dx} = c\frac{d\mathbf{p}}{dx} \qquad (4.9)$$

This derivative along the direction of motion with a particle having a velocity c can be integrated along the line yielding a constant of integration D. If the particle is massless, then c is the constant velocity of the particle.

$$c \int \frac{d\mathbf{p}}{dx} dx = c \, \mathbf{p} = D \ \ is \ \ a \ \ constant \qquad (4.10)$$

However, if the particle has mass, then c is replaced by the velocity v which is $\frac{p}{m}$:

$$\frac{p^2}{M} = D \qquad (4.11)$$

Now D can be selected to satisfy a convention for the definition of kinetic energy and in that case $D = 2T$. For a free particle with mass in a constant potential region the above equation is now satisfied. It is useful to see how the mass-free relations are derived verse the case of the particle with mass. The detailed derivation for a particle with and without mass are shown side by side in Table 5.

The massless column produces a first order differential equation for a massless particle. This equation is usually augmented by the self energy for starting the derivation of the Dirac equation. When looking at an individual particle, a photon, pAxion or electron at low energy or the lowest energy possible, the mass term in the above relation can be a constant absorbed into the energy or zero. The absorbed constant could represent the self energy if that serves a purpose and is at the total discretion of the user.

$$cp = E \ and \ cp = -E \qquad (4.12)$$

In the case of the mass-free derivation, the self energy of the particle, or the mass dependence can be buried in the energy term. Because this term is a constant, we have the latitude to have the self energy subtracted from the total energy, resulting in our selected constant. Utilizing the positive energy solution with an embedded self energy as:

$$cp = E - E_{self} \qquad (4.13)$$

Rewriting the equation we have:

$$cp + E_{self} = E \qquad (4.14)$$

The above equation is the starting point for the derivation of the Dirac equation. The main feature is that the self energy contains a mass term that is explicitly added to satisfy the conservation of energy and it is in the same form as the Einstein mass energy relationship. The same can be done for the negative energy solutions without taking the square root of the relativistic mass, momentum and energy relation. For a massless particle, the self energy of the particle drops out. The total energy, $E = T + V$, is just the sum of the kinetic and potential energies. So the above equation can be written in a second way with no external potentials just as:

$$T = E \qquad (4.15)$$

From the far column for a particle with mass we have the relation $\frac{p^2}{M} = D$. The quantity D is taken as 2T and once the potential energy is added the equation can be rewritten as:

$$\frac{p^2}{2M} + V = E \qquad (4.16)$$

The parallel derivation of the equations is summarized in table 4.

Table 4: Derivation Paths for the Quantum Equations

step	M = 0	M ≠ 0 Schrödinger	M ≠ 0 Dirac
$\frac{dp}{dt} =$ *3rd Law*	0	0	0
$\frac{dx}{dt}\frac{dp}{dx} =$	0	0	0
$\frac{dx}{dt} =$	c	v	c
velocity $\int \frac{dp}{dx} dx =$	constant	constant	constant
D constant *eg.* $\pm E$ *or* 2T	$cp = D$	$vp = D$	$cp = D$
set constants	$cp = E$	$\frac{p^2}{m} = 2T$	$cp = \pm E - mc^2$
add potential *energy*	$V = 0$ for now	$+V = V$	$V = 0$
result	$\mathbf{cp = E}$	$\frac{\mathbf{p}^2}{\mathbf{2M}} + \mathbf{V} = \mathbf{E}$	$cp + mc^2 = E$ $cp + mc^2 = -E$
add random *walk*	$\frac{ic\epsilon}{\hbar}(\mathbf{p}^2)_r + \mathbf{cp} = \mathbf{E}$	– ×	×

The equation with mass at low energies becomes the Schrödinger equation in free space when the kinetic energy is replaced by the squared momentum operator divided by mass and the energy operator is applied on the right side. The precursors of both the Dirac and the Schrödinger equations come directly from Newton's third law when considered in a space with no applied forces on the particle. These equations represent two equivalent forms used in different energy domains. The equation 4.14 containing the self energy, when made to conform to special relativity , produces the Dirac equation and the second form produces the Schrödinger equation. The derivation starts for the new wave equation with a massless particle equation, with the aim of describing the particle's rest state with mass. The reason is simply that mass cannot be assumed to be an a priori property of a particle.

The solutions of the mass-free first order equations are simple in the case when $E = 0$ in Cartesian coordinates. The solution is a constant and in spherical coordinates it is $-\frac{1}{r}$ and when normalized and multiplied by the spherical coordinates Jacobian also produces a constant that can only be normalized within a finite volume. That is to say, a particle with zero energy has a uniform probability of being found over all space. For this statement to be realistic for some particle, a limiting volume for the particle contained must be specified. This second requirement for a defined particle volume is unrealistic. This non-physical result is unacceptable for a physical particle at rest as it is defined only to be everywhere.

Summary

Starting with Newton's third law the form of the two basic equations of quantum mechanics can be isolated. The Schrödinder equation form only requires selecting a specific integration constant which is allowed. The massless first order wave equation provides a good starting point for the Dirac equation selecting a different integration constant where mass is embedded in a self energy term. The massless equation as a starting point is even more important for deriving the property of mass by allowing this first order equation to be expanded to include the interactions with the vacuum state or a solid state medium. These collective random interaction with the vacuum state result in an expression that explicitly matches the characteristics of mass that were exposed in solving the simple potential problems.

4.4   Mass Equation

The two examples of the effects of random processes on bosons and fermions, can be dealt with in two ways. First starting with an intrinsic naked mass then enumerating all the interactions as done in quantum field theory to compute the actual mass. This requires knowledge of all the detailed interactions and the validity that there is an intrinsic naked mass. More simply it is possible to deal with the interactions collectively. The latter approach will be used in this section. The algebra required for the wave functions comes from the standard development of quantum mechanics [30], which requires the norm of the wave function to represent the probability density distribution of the particle. The momentum operator being a spatial gradient allows this simple dispersion relationship to be transformed into a linear equation, where $u(x)$ is our wave function and b is a constant proportional to energy in the time independent zero mass wave equation previously derived.

$$u'(\mathbf{x}) = bu(\mathbf{x}) \qquad (4.17)$$

This is the simplest form of the quantum wave equation using the linear dependence of momentum to energy derived just above. But it is rather too primitive because it shows that change in the function is dependent upon the instantaneous value of the function. The concept of how a particle interacts with space in the above relationship indicates a prompt response is compromised by the discreet nature of the particle-space interactions. What has been shown is that the boson polarization and the fermion interaction with the vacuum respectively produces a phase delay for attraction or a local dynamic displacement representing repulsion. The properties are dependent on the random mean displacement ϵ caused by vacuum interactions or exciton scattering in the solid state. This response can be expressed as dependent on the function in its initial position, which then suffers a displacement of ϵ. So a more accurate expression for the wave function relation is:

$$u'(\mathbf{x}) = bu(\mathbf{x} - \epsilon) \qquad (4.18)$$

This transforms to:

$$u'(\mathbf{x} + \epsilon) = bu(\mathbf{x}) \qquad (4.19)$$

and then in a Taylor expansion yields:

$$\epsilon u''(\mathbf{x}) + u'(\mathbf{x}) = bu(\mathbf{x}) \qquad (4.20)$$

The quantity $\epsilon > 0$ was chosen initially as a positive scaler. It represented a delay via a displacement by being subtracted from the local position, but in the transformation it ends by scaling the second derivative as a positive constant. It is from this quantity that mass becomes a positive quantity. If space is isotropic in terms of the interactions with the vacuum state, then the quantity ϵ and mass are simple scalers. Whereas, in the solid state where the isotropy of space is lost, the effective mass must be represented by tensor.

Including the displacement to model the particle interaction with the vacuum state makes the resulting equation different from both the Schrödinger equation and the basis for the Dirac equation because both equations describe a promptly responding system.

It is useful to look at the possible solution space that is actually covered by the two standard quantum mechanical equations. A simple figure that exhibits the constrained nature of the development of quantum mechanics is a plot of the allowed solutions in the effective momentum spaces. The allowed solution space in figure 4.7 is basically empty except the four lines with some sample bound states shown as black dots and with an origin that is not a valid state. This is either an incredibly frugal use of possible solution space or we are missing a description that extends measurable activity into the unused regions. In analogous electromagnetic propagation descriptions, in some materials there are regions that are not classically allowed, but they are not restricted to four lines and some isolated points. When quantum behavior is added to the possible classical electromagnetic solutions, the allowed space becomes greatly expanded. The most important element in this diagram is the missing origin, no physical quantum mechanical description of a massive particle at rest.

The Taylor expansion then contains the structure required as a source for the new wave equation that lead to something different than the Schrödinger or Dirac equations. The magnitude of the delayed displacement ϵ is inversely proportional to mass in this analogy. The quantum Brownian dynamics yields two results: The first is the mass independent of the sign of the particle-space interaction and second is the requirement for a probabilistic measurement space that is described by our current understanding and

Figure 4.7: *Allowed eigenfunction propagator exponent solution space for bound and free particles*

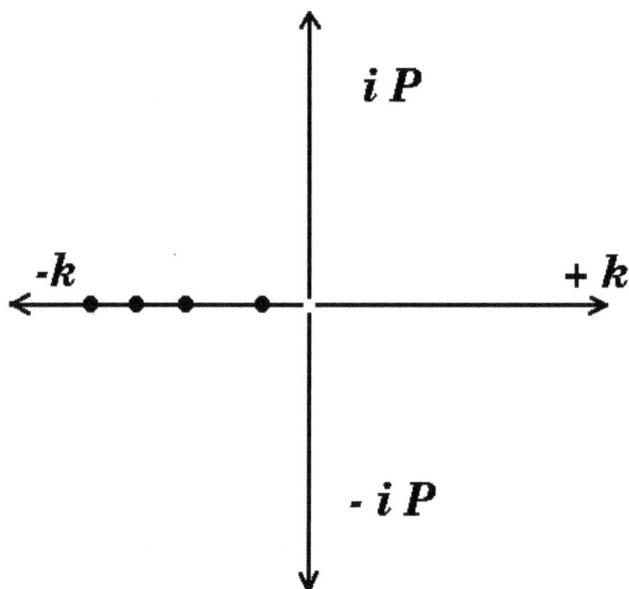

Sparse
Propagator solution space are
the 4 lines without the center

algebra of quantum mechanics. Since ϵ is a mean value of a random variable that describes either the net effect of a phase delay or a physical displacement, the new equation's solution is directly based on this random variable. This activity does something else: By including the randomizing effects of the vacuum interactions, space for the particle has taken on completely spherical symmetry. We don't know if from our measurement perspective that such a symmetry existed before considering the randomizing effects of the interactions or whether this symmetry is only due to the information we gain from specific particle's interactions for which detailed properties are known. All we can say is that when particle motions can be randomized with either phase delays or displacements we expect their properties to be isotropic and in particular they now possess a mass. This, however, says nothing of the underlying symmetries of the space.

4.5   The Rest State

By including the random processes of the particles interactions with the medium, the problem has been bootstrapped into a description that avoids summing over all possible massless interactions. Putting the Taylor expanded equation into a form that can be solved is required. Because ϵ is the mean of the random variable that resets the location of our particle, it appears to be most easily treated in Cartesian coordinates where the random variable has three vector components ϵ_x, ϵ_y and ϵ_z. The result of the displacement in the three spatial random variables. Where $\bar{\epsilon} = \epsilon_x \hat{i} + \epsilon_y \hat{j} + \epsilon_z \hat{k}$

$$u'(\mathbf{x} + \bar{\epsilon}) = \frac{\partial u}{\partial x} + \frac{\partial u}{\partial y} + \frac{\partial u}{\partial z} + \langle \epsilon \rangle \nabla^2 u \qquad (4.21)$$

The momentum operator, which is the source of the derivative, is really a scaled derivative and that scaling is in place prior to the Taylor expansion to maintain a correct set of units. So the units must be put in order:

$$\mathbf{x} = \frac{i\mathbf{x}'}{\hbar} \qquad (4.22)$$

$$\langle \epsilon \rangle = \frac{i \langle \epsilon' \rangle}{\hbar} \qquad (4.23)$$

Where the primed terms have units of meters. Now rewriting the Taylor expanded equation in the primed units and then dropping the primes leaves:

$$\frac{c\hbar}{i}\left\{\left(\frac{\partial u}{\partial x} + \frac{\partial u}{\partial y} + \frac{\partial u}{\partial z}\right) + \langle\epsilon\rangle\nabla^2 u\right\} = Eu \qquad (4.24)$$

The first inclination is to transform this to spherical coordinates and separate the angular variables from the radial variable, however, the first order derivative makes that difficult since there is a factor of $\frac{1}{r}$ in the terms of the angular derivative and a factor of $\frac{1}{r^2}$ in the second partial derivate of the angular terms. This means that the variables are not separable. But more importantly, the nature of the equation has changed. This equation is a product of the random behavior of the particle via its interaction with vacuum or a medium. The particle will be taking a random walk as described by the equation. Though we can specify angular coordinates, there are no references and no symmetry is expected to be fundamental to this behavior. That is not to say that a symmetry cannot emerge from the details of the behavior, but to the first order the random behavior will destroy any angular symmetries.

$$u(r, \theta, \phi) = u(r) \neq R(r)Y(\theta, \phi) \qquad (4.25)$$

This simplifies this hybrid equation that has a statistical operator and a propagation operator by forcing the dimensionality to just one variable and spherical symmetry. In spherical coordinates the equation becomes.

$$u''(r) + \left(\frac{2}{r} + \frac{1}{\langle\epsilon\rangle}\right)u'(r) - \frac{iE}{\langle\epsilon\rangle c\hbar}u(r) = 0 \qquad (4.26)$$

This equation does not really look much like the Schrödinger or the Dirac equation because it is going to map out a solution space that is complementary to those two equations. It is also operating at a level that is lower than either of the two standard equations by providing a description of mass.

$$\langle\epsilon\rangle = \frac{\hbar}{cm} \qquad (4.27)$$

The new quantum wave equation now has the mass of the particle self consistently included. By rewriting the constants in terms of the curvature of $\kappa = \frac{1}{\langle\epsilon\rangle}$ the equation is put in a form where the last term has the ratio of total energy to self energy.

$$u''(r) + (\frac{2}{r} + \kappa)u'(r) - i\kappa^2\frac{E}{mc^2}u(r) = 0 \qquad (4.28)$$

What stands out in this differential equation is that the energy is scaled by the Einstein self energy relation mc^2. The derivation did not involve the use of the Lorentz transformations, only the condition derived in chapter 2 defining a scale dependence on mass which is of quantum mechanical origin generated by random displacements. This feature assured the solutions to this differential would be interesting if not exceedingly useful.

This equation does not have a simple Bessel function solution because of the second term in the first derivative product of the wave equation. It, however, does have a solution expressible in terms of some common functions. Taking $\alpha = \kappa^2\frac{E}{mc^2}$ results in two solutions where F1(a,b,c) and U(a,b,c) are the confluent hypergeometric functions (see Appendix B) and p and q are constants then:

$$\gamma = 1 + \frac{\kappa}{\sqrt{\kappa^2 + 4i\alpha}} \qquad (4.29)$$

$$u(r) = e^{-\frac{1}{2}(\kappa+\sqrt{\kappa^2+4i\alpha})r}$$
$$\times \{pU(\gamma, 2, \sqrt{\kappa^2 + 4i\alpha}r) + qF1(\gamma, 2, \sqrt{\kappa^2 + 4i\alpha}r)\} \qquad (4.30)$$

The first thing to notice is the terms in α which will be complex when $E \neq 0$ and $m \neq 0$. In fact for this equation $m > 0$ becomes a requirement. In the sparse solution space of figure 4.7 this would allow a solution to cover a portion of the empty space because the arguments of the terms are of the form $(a + ib)r$. When the time dependent portion of the wave function is multiplied with $u(r)$ the complex terms in the exponent allow the rest state mass function to translate through space.

A key solution of this equation will be when $E = 0$, rest state. The solutions in the U confluent hypergeometric function looks like a simple hyperbola and it produces a finite integral for the wave function integrated over all space much like the solution of the Schrödinger and Dirac equations. The F1(a,b,c) is not a bounded function (increases with r) and does not appear to be a physical solution in an unbounded region. When $E = 0$, F1 has a value of 1 over all space. The details of the U and F1 confluent hypergeometric functions are found in appendix B. These solutions both appear to be valid and interesting. We will examine the U solutions which will produce a finite probability in an unbounded space. However, in a space with obstacles and boundaries the F1 form of the solutions will be used to determine the probability distributions. For the case when $E = 0$ the resulting solution for the U form appears below:

$$u(r) = pe^{-\kappa r}U(2, 2, \kappa r) \quad for \quad E = 0 \tag{4.31}$$

The hypergeometric function U falls off rapidly at the radius determined by the mass relationship. Within the radius $r < \epsilon$ or in the plot $\kappa r < 1$ for the boson, the wave function is determined by the mass free quantum wave equation and at zero energy it is a constant resulting in a scaled density function of a particle as shown below:

The function in figure 4.8 when it is integrated over all space, produces a finite result that when normalized is independent of E. This is an important feature in defining the particle, in that the normalized integral of the density is a real constant and density distribution is also invariant, the same in any energy state as it is in the rest state. Since the raw wave function is a hyperbolic function, it is easier plotted on a Log_{10} plot to view its behavior over a wider range. The wave function is shown below:

When $E \ll mc^2$ the resulting solution is:

$$u(r) = e^{-\kappa r}e^{-i\frac{\kappa E r}{mc^2}}$$
$$\times \{pU(2(1 + i\frac{E}{mc^2}), 2, \kappa(1 + i\frac{2E}{mc^2})r)) \tag{4.32}$$

Figure 4.8: *Scaled probability density function for a particle with mass that is invariant with respect to energy for both $E = 0$ and $E = mc^2$ which map to the same function.*

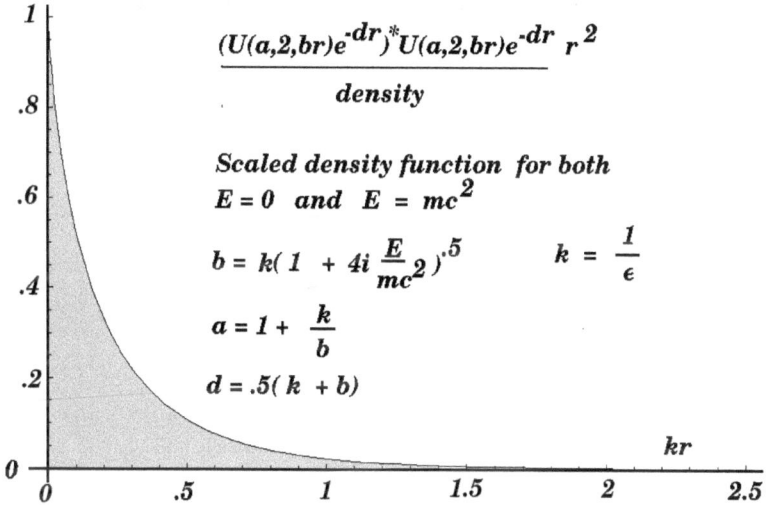

$$\frac{(U(a,2,br)e^{-dr})^* U(a,2,br)e^{-dr} \; r^2}{density}$$

Scaled density function for both
$E = 0$ *and* $E = mc^2$

$b = k(1 + 4i\frac{E}{mc^2})^{.5}$ $k = \frac{1}{\epsilon}$

$a = 1 + \frac{k}{b}$

$d = .5(k + b)$

Where the propagation vector for the particle with finite energy becomes $\frac{\kappa E}{mc^2}$. If the ratio of $\frac{E}{mc^2} = 1$ and the wave function $u(r)$ is normalized and plotted against the zero energy function a rather important result becomes visible. The scaled density plot of this function is not much different than the previous function and generates the same scaled density figure as $E = 0$ function. The second interesting feature of this function is that the factor of multiplying by r is complex and opens up the unused territory in the solution space for quantum particles. There is now a wave function which takes advantage of the full propagation space and mass is a natural feature generated from the derivation of the equation that produces the wave function. The coordinate representation for the above equation is strictly in the frame of reference of the particle itself. All relativity considerations are satisfied by this fact alone. The other feature is that all energies are scaled to the self energy of the particle which is solely determined by the interactions with the vacuum state whether the particle is a boson or

Figure 4.9: **Log base 10 plot of the radial dependence of the wave function for E=0 from the new wave equation**

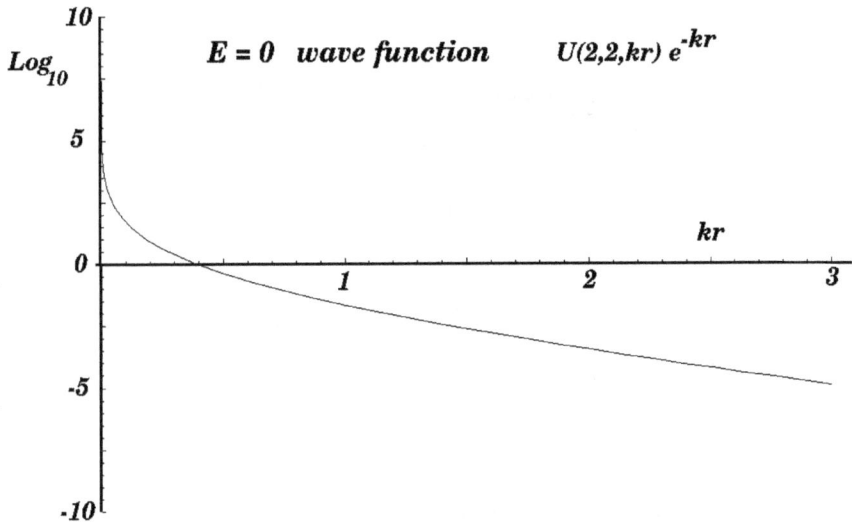

a fermion. The self energy arises naturally without requiring the time dependent Lorentz transformation derivation.

The function above $u(r)$ can be expressed as $u(r, E)$ where E is a variable. The scaled density function $\rho(r, E)$ is defined as:

$$\rho(r, E) = \frac{u(r, E)^* u(r, E)}{4\pi \int_0^\infty u(r, E)^* u(r, E) r^2 dr} \tag{4.33}$$

This normalized density function $\rho(r, E)$, at least for values of $\frac{E}{mc^2} < 10$, tested numerically generate the same function independent of E. This can be seen the exponential term $e^{-\kappa r}$ in the density function which is real and is independent of energy in the non-relativistic case. This is about as close as one can come to a condition of mass quantization that the density distribution of the particle is an invariant as a function of the energy of the particle. The effects of relativity are automatically included because the mass scales as the curvature which allows a proper transformation of the new wave equation second derivative term. The result of considering a quantum Brownian behavior generates a mechanism by which only a single property value, mass, for a particle can be assigned to the particle and over a range of energies and remain invariant. This is not so different from how the quantization of angular momentum or spin is a product of the requirements of what constitutes a proper solution of their specific wave equations.

4.6   Deriving the Repulsive Fermion Potential

We have been guessing the form of fermion potential interaction with vacuum. A detailed knowledge of the repulsive potential is needed before arguments about the mechanism defining a fermion's mass can be completed. The first application of the rest state solution is to generate the repulsive relaxation potential when an electron encounters a virtual electron-positron pair. This modeling takes the density distribution and relaxes the position of the three particles symmetrically from the positive particle that remains central. Using a Monte Carlo integration to compute the potential affecting the initial electron allows a picture of the repulsive potential to be generated. An interesting feature of this integration is that infinites due to the collocation of two particles on the same site can simply force a local finer discretization and random site selection until the collocation no longer occurs. This is only possible because of the inherent mass distribution is no longer a point at this level of analysis.

For a set of three point charges represented by rest state charge distribution the potential generated for a symmetric displacement of x of the two like charges from the center yields $V(x = 0) = 0$ and $V(x) \sim -\frac{1}{2x}$ for $x > 0$, which has a singularity at zero that acts like a repulsive pole. This repulsive singularity at the origin is the only means by which the repulsive potential can generate a mass at all length scales for a particle such as an electron. Functionally in Cartesian coordinates this can be represented as a product of the unit function [44]:

$$U(x) = 1 \; for \; x > 0$$
$$U(x) = \frac{1}{2} \; for \; x = 0 \qquad (4.34)$$
$$U(x) = 0 \; for \; x < 0$$

The potential pole at the origin becomes

$$V(0) = -4U(0)U_{inv.}(0) \qquad (4.35)$$

Where $U_{inv.}(x)$ is:

$$U_{inv}(x) = 0 \; for \; x > 0$$
$$U_{inv}(x) = \frac{1}{2} \; for \; x = 0 \qquad (4.36)$$
$$U_{inv}(x) = 1 \; for \; x < 0$$

A first guess at the total potential is then:

$$V = 4U(\mathbf{x} - \mathbf{x}_o)U_{inv}(\mathbf{x} - \mathbf{x}_o) + f(\mathbf{x} - \mathbf{x}_o) \qquad (4.37)$$

Where $f(\mathbf{x} - \mathbf{x}_o)$ is the potential interaction defined everywhere except at $\mathbf{x} - \mathbf{x}_o$. In this case for the Coulomb potential it is the function $\sim \frac{1}{\mathbf{x} - \mathbf{x}_o}$. It is the unit function's more common relation Dirac delta function which is obtained by differentiating the unit function. The first term of the potential generates no force at $\mathbf{x}_o$ when interacting with a test charge at that location.

In the Monte Carlo calculation (the code is listed in Appendix B) three particles with the ground state wave functions just generated for the $E = 0$ state are used. When the three particles are superimposed at the origin the averaged potential interaction remains zero, with a high variance in the Monte Carlo calculation. This variance in the calculation actually captures the random features of the physics of the interaction that drives the vacuum interaction process. However, as the polarization drives the two negative electrons apart a potential well develops around the origin. So that the prompt initial interaction when the three wave functions overlap is a net repulsive potential even though the net additional charge is zero. The energy that the electron gains in occupying this effective potential is used to power the magnetic spin field of the electron. A schematic of the process is shown below:

Figure 4.10: *Snap shot of the relaxation dynamics of the electron vacuum interaction with an electron-positron pair.*

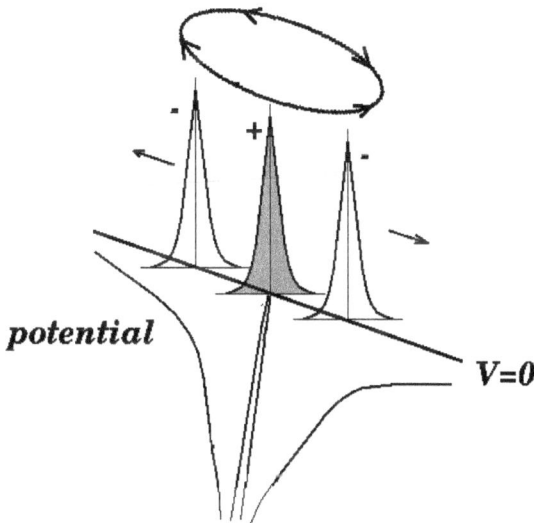

two dimensional section of the 3-d spherical potential well generated by relaxing the position of two electrons about a positron

The deep one over r charge potential supplies the electron energy to trigger a three particle interaction that generates its dynamic properties. So in the case of an elementary charge particle like the electron interacting with vacuum it becomes a pseudo-3 particle problem. In terms of the electrostatic potential field, there is no additional long range effect except a randomness with variance of ϵ over the interaction distance. The above model is symmetric, which is naturally unstable, with the particle aligned and that generates a minimum potential function, but the particle positions should be random and that deviation from symmetry generates a chase condition that will drive the potential back to its minimum and also generates a two particle rotation. Once energy is in the spin field there is an inherent memory for the state with the field sustaining the new lower symmetry state. This then reduces the spherical symmetry the problem initially possessed. Even though the model is symmetric, the long range nature of the wave function and the sampled local interaction requires a large number of potential samples to be calculated to reveal the potential function. Even with 10^7 samples for each displacement in the following figure there is a large variance in the result. This can be seen particularly for the more than 15% of the samples end up with an estimated positive potential interaction.

Figure 4.11: *Computed three particle interaction showing the random nature of the process and the resultant effective potential well. The continuous curve was fitted to the data is* $v = -\frac{2.42 \times 10^8}{r^2} - \frac{7.18 \times 10^8}{r}$.

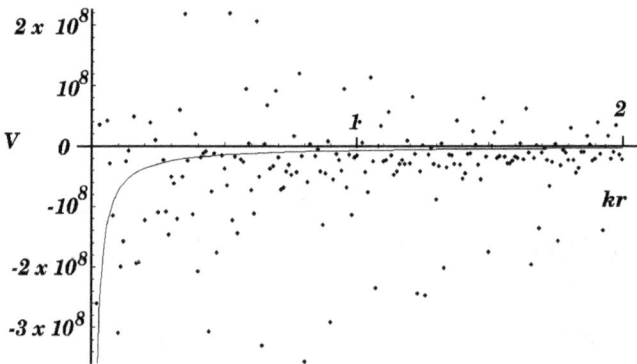

Pauli's pronouncement [45] that spin has no classical counter part does not limit the application of such a potential in finding a self consistent quantum description of the two possible spin states from this many-body model of the interaction of the electron.

Summary

This section is based on considering a series of events to modify the basic quantum description to define the local mechanics of the particle under study. This came about by looking at two very different forms of random quantum activity that bosons and fermions would suffer. This random activity is the key to defining the properties that we measure depending on the scale at which we are looking. This change in view point results in a new quantum wave equation located with the particle of interest. A wave equation exists that describes the rest state of a particle with mass. The solution is a well-behaved description of the total particle density distribution that is invariant with energy, making the non relativistic mass invariant with energy. The integral of the density function generated from these wave functions are finite when integrated over all spaces, consistent with a valid particles existence. The valid rest state is a feature not found for the $E = 0$ solutions of either the free particle Schrödinger or Dirac equations, which are not bounded. This reapplication of the Brownian like activity on a reduced scale can now be used to define the origins of mass. As a device it shows that the vacuum state is an active and interesting place that controls some of the most basic properties of particles including their mass. The theoretical result that is most significant is that it extends Newton's first law to quantum mechanical particles by allowing for a meaningful definition of the rest frame for a quantum particle and its wave function.

Table 5: Free Particle Quantum Equations

equation	solution	rest state
$-\frac{\hbar^2}{2m}\nabla^2 u = i\hbar\frac{\partial u}{\partial t}$	$e^{i(px-\frac{Et}{\hbar})}$	ambiguous
$(c\alpha \bullet \mathbf{p} + \beta mc^2)\psi = E\psi$	array $\times e^{i(\mathbf{k}\bullet\mathbf{r}-\omega t)}$	ambiguous
$\epsilon u''(\mathbf{x}) + u'(\mathbf{x}) = Eu(\mathbf{x})$	$e^{(a+ib)\frac{r}{\epsilon}}U$	valid state

4.7　Particle Volume

The property of the wave function $u(x)$ in accommodating a displacement as a phase shift

$$u(x) \longrightarrow u(x)e^{-i\theta} \tag{4.38}$$

on a route of suffering periodic random phase delays represented in the variable θ where

$$\theta = k\Delta x \tag{4.39}$$

is a feature that can only occur if the principal description of a particle or a field were not a real function. Where k is the propagation vector. The measurable state, the norm of the wave function, allows the phase to play a roll if it becomes a random variable.

Figure 4.12: **Schematic of a particle interaction resulting in a random phase delay**

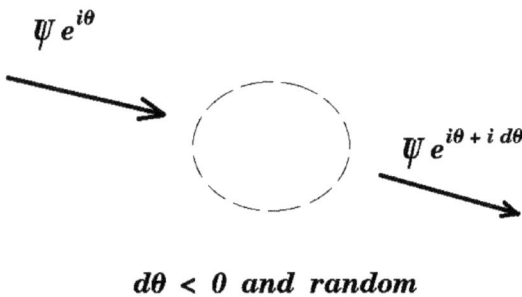

$\Psi e^{i\theta}$

$\Psi e^{i\theta + i\, d\theta}$

$d\theta < 0$ and random

The role the phase can play is to introduce an effective random displacement, while energy is conserved. The result is the mean displacements ends up being inversely proportional to the mass of the particle. The net effect in these activities results in the self energy for the particle being transformed into the role of the particles mass. If the phase is constant and not a random variable, then the particle is massless. This description changes the calculation of dealing with the specific potentials involved with a particle interacting with the vacuum state and replaces it with an inertial mass and its dynamics.

This random walk of a quantum particle is created by the dense interaction with particle and anti-particles pairs. This process has characteristics different from a lattice random walk, with the effects collecting and determining the base dynamics. The displacement caused by this walk will be a measurable feature of the dynamics of the particle only as an average characteristic of the particle and that will be reflected in its mass.

Figure 4.13: *Dense series of fermion vacuum interactions resulting in a random walk*

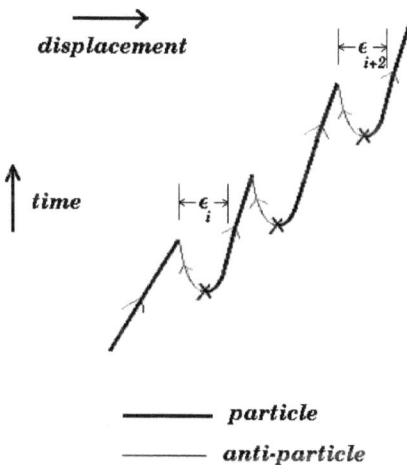

Since this is a random process and described as an event, any temptation to treat such a process by considering a gauge transformation effect on the wave function would fail because there is no continuously differentiable field that will accommodate the events.

The mass of both the boson and fermion can be related to the inverse of scale parameter, which is characteristic of the curvature of the volume of space required for the processes of the quantum Brownian motion. This correspondence is not to volume of the space, but to the inverse of the radius of that volume. A feature of these interactions for the fermion is that any two particles can interact within a plane, but the addition of a third particle requires a full three dimensional description. This gives the entire interaction an effective volume. Even if the particle descriptions are point-like, the net effect for a coarse observer is that there is a finite volume over which the action takes place. For the boson or photon, the interaction also involves a collection of particles and fields greater than two, which produces an interaction over a volume. So that even with point particles, their effective interactions can define a volume. However, the parameter coupling the interaction to mass or the self energy is the curvature, which characterizes an interaction volume rather than a point.

From a physical point of view curvature in itself is not a concept that stands on its own in a dynamic description of forces acting on the vacuum. The curvature can be considered a ratio between the area enclosing a spherical volume to the volume itself. When the factor representing area appears in a non static description, it is usually associated with the flux of a field that drives an interaction.

$$\kappa = \frac{Area}{Volume} = \frac{4\pi r^2}{\frac{4}{3}\pi r^3} = \frac{Flux}{Volume} = \frac{3\ FluxDensity}{r} \qquad (4.40)$$

With $Flux = FluxDensity \times Area$, the *Flux Density* as a local per unit area density, it must be a unit-less constant. In our representation of a quantum wave equation which gives a description of the rest state there are no other units attached to the curvature except the unit of inverse distance. That implies the *Flux Density* is unit-less. Being unit-less implies the *Flux Density* is a measure of space itself. Considering how the $\frac{1}{r}$ Coulomb potential can trigger a strong vacuum interaction in the space for itself is not difficult because of the unlimited depth of the potential. This makes the measure of curvature being proportional to the strength of the local *Flux Density* not seem irregular. Its value only has meaning when there is a finite curvature associated with a mass. This is an extremely simple statement with no qualifiers [46] and something one would expect to feature in the basic description of mass. So the perspective of curvature representing a local flux generated on the space is not excluded from this concept of mass.

Summary

From the particle point of view the curvature parameter that results is not felt to define the basic nature of the random interactions. The curvature parameter is just the ratio of the surface area of the interacting region to the volume of that region. Since the effects are centered in the frame of the particle and have no memory it is not a viscous drag. In the limit of the Dirac equation, the vacuum interaction effect is encapsulated into the self energy term mc^2 whereas in the Schrödinger equation it is simply ignored as a self energy term and a ballistic quantum description of a classical like particle is computed. These are all approximate models that give good service in their specific regions of application.

4.8   Algebra that Allows Mass

The algebra of the wave function contains the mechanics that allows mass to be defined as a quantum characteristic. This characteristic is embedded in the phase of the wave function. If the phase is not a random variable, then the operation is on the level where the interaction with space is either not important or being ignored. If a fraction of the phase is operating as a random variable because of the particle's interaction with the vacuum state, then the particle is acquiring mass. For a boson, the effect of having a random variable that can represent a phase delay also characterizes an attractive interaction. Similarly, the phase delay can represent a displacement that can occur and energy has been conserved with no change in the final state of the particle except for the displacement. The fact that the wave function can be represented by a complex function whose total measurable density is the product of the function times its complex conjugate that is invariant with respect to a displacement. This feature of the algebra allows the use of this very non-classical mechanism to generate mass. Newtonian mechanics and the general theory of relativity do not include a description of mechanics on this level. Therefore they cannot generate a mechanism to allow a fundamental description of mass. This description of mass automatically introduces a hierarchy into the way the problems have to be solved, particularly for bosons where on a small scale it leads to massless behavior.

There is a question that can be asked, and that question is whether all mass is generated by random processes for a particular particle? Could there be a fraction of the mass of the particle that is the intrinsic point mass of classical mechanics? An argument can be made that is close to a proof to challenge this point. In the previous integration of the Newton's third law one either has a particle with a constant velocity or an energy dependent velocity. The dynamic equations split at that point and the mass free equation cannot be merged with the Schrödinger equation into a composite equation because it makes the dynamics meaningless. The precursor to the Dirac equation contains the explicit inclusion of the self energy term with mass, but this yields no rest state solutions in its final form. If the only description that exists contains intrinsic mass then that description would violate the extension of Newton's first law to quantum particles as there would be no rest state. It, therefore, appears that the quantum random motion via its interaction with the vacuum state or medium is the requirement for an elementary particle to generate mass starting with a mass free description of a particle. If the particle requires an intrinsic mass to be modeled it is not the most elementary description of the particle. The connection between mass and energy from this point of view appears not as an arbitrary definition, but rather as an expression of the strength of the local interaction between the particle and the vacuum state. The resultant scaling of the particles energy with the Einstein mass-energy relation is a natural consequence of this basic interaction that generates mass.

One thing that has been ignored in this description of mass is a discussion of the scale on which the new mass equation is applied. For the lightest charged fermion, the electron, the scale at which the wave function has meaning is on the order of 10^{-15} meters. We will consider some very light bosons and the scale of the wave function will increase significantly. In terms of solving problem on a much larger scale such as the hydrogen atom, the wave function of the electron can be approximated as a point mass.

The selection of a particular wave equation for the analysis of a problem is determined by the problem under considerations. The development of the rest state wave equation and its solutions supplies some meaning to the concept of mass and the particles distribution in the rest state. If the charge distribution mirrors the mass distribution in the rest state there is simplification in doing potential calculations since the point particle potential infinities disappear at this scale. For example in doing an atomic scale problem the scale of the rest state of a electron will be much smaller than the scale of an atomic problem as a whole. This allows the use of the mass parameter for a material description in the atomic problem with a simple coordinate representation found in the Schrödinger equation or the Dirac equation. So the new quantum wave equation operates at one level lower on the interaction scale than the other two equations. This similar to the analogy already introduced of thermodynamics setting on the foundation of statistical mechanics.

If multi-particle wave functions are considered the application of the new wave equation on the total wave function of an ensemble of particles may give some insight into the behavior and limits to entanglement and the dynamics of the system. The scale over which the total wave function will operate will be determined by the strength of the interaction of the particles with their environment and thus describe the limits to the valid use of the multi-particle wave function. The description breaks down into individual components, first will be the center of mass behavior of the entire function in coordinates of $\frac{r_1+r_2\cdots r_n}{n}$ verses the individually coordinates referenced to the center of mass. This dual description holds until the collective state is no longer a physical descriptions because of interaction with the environment and the single particle descriptions become the only remaining description.

Summary

The complex algebra of the wave function, which can possess a phase that can be used in part as a random variable, provides the mathematical mechanism to describe mass for a variety of particles using different processes to interact with the vacuum state. The arguments of possible interactions for boson's and fermion's are only intended to show that mass can be defined by such processes alone. Intrinsic classical mass does not appear to be possible because its description does not allow a rest state that would violate Newton's first law.

Introducing the interaction volume for a particle yields a new hybrid quantum wave equation which gains spherical symmetry due to the random processes affecting the particle it describes. The solutions of this new wave equation are complimentary to the standard wave equations, by producing a description of the rest state. The solution of this new equation features a probability density function being invariant as a function of energy which yields a quantum description of mass that is only dependent on the basic random interactions the particle suffers. Its not the character of the interactions that are important it is only the net mean random displacement the particle suffers that determine the mass of the particle. This lays the foundation for the calculation of masses of individual particles from the details of their interactions within the medium they are embedded. Mass's quantum description now removes the problematic point-mass description of classical mechanics.

Chapter 5

Time, Measurement and Relativity

We have picked our way through a couple of derivations using time independent equations to produce a result for a particle's random interaction with the vacuum state that produces a result defining mass and the particle's self energy. This was on purpose because time dependent operators on this scale become a problem when dealing with an individual particle. Time is not involved in the measurement of a single event, time is only a computable quantity when measuring an interval. To measure an interval three things are needed: something to measure, a particle is the simplest object to make a measurement on and two events recorded by a second and third particle. This is a total of three real and separable objects. The interaction with the vacuum state for a free particle or a solid state excitation involves no extra second and third particles that are separably measurable to record the activity.

In physics time appears only to have a literal meaning [47]. It is this concept of time that sharpens the divide between the Dirac/Schrödinger equations and the new time independent description that generates mass. The nature of the momentum operator allows this separation of the problem to occur. In fact space-time (x, y, z, t) which is the normal macroscopic description gives way to a simple space description (x, y, z) when considering particles interaction solely with the vacuum states. This follows a very simple literal definition of time that originates with the minimum set of objects required to define an interval. The two recording particles constitute a primitive clock or record. This argument ties together the requirements of a measuring instrument and the measurement itself specifying the simplest measurement that can be made. Time requires a significant amount of physical

machinery in order to be made real. Whereas, on the level of the
particle alone, the dynamics of the problem is removed by being
in the same frame of reference of the particle. Energy in that case
only appears as a scale factor in the particle's wave function. Being
in the frame of reference of the particle, removes dynamic considerations, time and relativity considerations. If a physical current
is to be studied then it must be modeled with time, however, a
randomly generated particle in a diffraction apparatus need not
be modeled as a current but can be analyzed simply as a particle
in a bounded region of space.

This literal definition of time then explains something about
where special relativity can be applied. It can be applied anywhere
time can be measured and does not apply anywhere time cannot
be measured. Therefore a knowledge of time is something that is
not needed for the concept of mass[1] or for an elementary particle
to exist. A composite particle with three components crosses this
boundary only when the individual components can be separately
measured. This seems to be a sufficient condition to allow nucleons
to be treated as elementary particles in their frame of reference.
This is a strong statement for asymptotic freedom for the components of the nucleons or any multi-component elementary particle.
This viewpoint separates measurements into two classes, those in-
volving time verses those that are time independent. This notion
has application in the simple double slit diffraction of particles
when there is only a single particle in the apparatus. Incorpo-
rating random behavior in the second order term of the new wave
equation replaces any explicit time dependence and the mechanism
of time with a history of random displacements.

$$\epsilon u''(\mathbf{x}) + u'(\mathbf{x}) = E u(\mathbf{x}) \qquad (5.1)$$

This process divides the two domains where the individual
models maybe applied and directly ties the concept of measurement to the observation of quantum behavior.

---

[1]The Clay Institute has sponsored a prize for defining particle mass from a
4-D field theory. The challenge was questionable as it appears to be possible
only in 3 dimensions or less where they were seeking 5 dimensions or more.

Chapter 6

Examples

Typically mass is treated as a renormalizable scale parameter and as an appendix to the standard model [48] that is needed to understand the operation of the weak forces in nuclear decay. There is a recent history that gives a chronology to the particular contributions and advances in this area by Baggott [49]. This is a complicated place to begin to deal with the origination of mass. Mass is dealt with as an intrinsically rescale-able property, a parameter, so that other properties conform to measurement. The reason for this approach using perturbation theory is there was not a consistent understanding of origin of mass for the electron or the photon. There were strong reasons to assumed the photon was massless at all scales. We have given some examples of how interactions of the photon and the electron in vacuum could possibly excite a random motion that would be sufficient to generate their respective masses. We were really not very concerned about the specific details at this point for these processes only their possible existence. If a process existed to drive this local random activity then its description would produce a wave function that would have a characteristic mass, dependent on the scale parameter of the random activity. So instead of looking at the nuclear origins of mass we will explore some low mass possibilities that operate on a much larger scale that will be easier to examine. We will start with the requirements of the zero mass state of the charge carriers in graphene. This will be followed by a look as some recent astrophysical data that can be used to compute the mass of a photon. Then the large scale magnetization waves in steel are considered in more detail which have a mass between that of the electron and the photon. Finally the data from supernova 1987a is reviewed to see if there are useful mechanisms to describe the source of the rings from ideas developed on the properties of low mass objects.

6.1 Zero Mass Excitations

The scaling minimum of the attractive potential for a boson provides a range where zero mass properties can be found. The case for a fermion is more complex. In order for fermions to have a mass at all scales, the local potential interaction with the space must also scale toward zero. If the repulsive core does not scale below a finite limit or the repulsive potential cannot be generated then the particle/exciton mass will not be defined below that scale nor will the term fermion have any meaning. It will simply be a particle.

Graphene is a model system [50], [51] of an exceedingly strong two-dimensional close-packed sheet, whose coupled electron-hole charge carriers in transport have a dispersion curve that is linear. Graphene is studied on a small scale where dimensions are measured in hundreds of microns with a low density of charge carries because of the short finite lifetime that is necessary to complete their travels. It is not apparent that in the environment of a strong solid with a reduced dimension, that the mobile charge carrier can efficiently polarize the structure to form a local potential interaction on a scale less than the dimension of the sample and thus gain mass. Graphene's charge carriers provides a second example of massless behavior in addition to the photon but on a smaller scale that is large by atomic standards. The high velocity of the carriers in graphene are limited by the finite stiffness of the lattice. The inability of the lattice to be strongly polarized by an individual component on a time scale where the charge carriers are defined over a finite area will generate no mass for either species. Because of their inherent high velocities, the minimum area for defining mass will be large. Currently this area is larger than the material samples measured. The fact that the carrier's dispersion curve reflects either a boson below the threshold to acquire mass or a particle that cannot excite a local repulsive interaction, implies the wave function of the carrier describes a simple very weakly interacting ballistic particle.

6.2 Photon

The case for a massless photon is often given by the argument that
in order to satisfy gauge invariance the photon must be massless.
This invariance must hold over all space so therefore the photon
is defined over all space. This is not a physically realistic argument. For this to be true, the wave function of the photon must
be computable over all space. Because from the rules of quantum
mechanics, the total probability of finding the photon must be
finite the integral over the photon density must be over a finite
volume of space or the result would be ambiguous both mathe-
matically and physically. To show this, only the representation of
the photon wave function is required. If we use the description
of a plane wave photon [52] with polarization ϵ^μ the four-vector
potential is equivalent to its wave function:

$$A^\mu(x;k) = \frac{\epsilon^\mu}{\sqrt{2kV}}(e^{-ikx} + e^{ikx}) \qquad (6.1)$$

The problem comes in with the term $\frac{1}{\sqrt{V}}$, which appears in
both the vector potential and the wave function. It is not a computable expression if the domain of the photon is infinite. A three
dimensional quantity that is infinite in each dimension is an un-
necessary complication. Therefore, the domain of the photon, over
which gauge invariance [13] and the massless character of the photon can be used, must be finite.

One of the first formal arguments for a self energy and mass of
the photon was made by Gregor Wentzel in 1948 [20], where the
estimate was a considerable fraction of the electronic rest mass.
Since then, those estimates have dropped to the point that made
designing laboratory experiments to extract the minimum self energy difficult. The net interaction of the vacuum in the form of
a weak phase retardation can give rise to mass. For the photon
at high energy, the dispersion relation is that of a massless particle because the self energy is so small. On a laboratory scale the
low energy photon can be taken to zero frequency to generate a
static electromagnetic field. This implies that the photon disper-

sion curve is a scale dependent property. On a small scale, the dispersion relation is just a linear relation between energy, E, and momentum $E = cp$.

Figure 6.1: *Photon dispersion relation in two measurement volumes, the one on the left on a small laboratory scale and the one on the right on a large astronomical scale.*

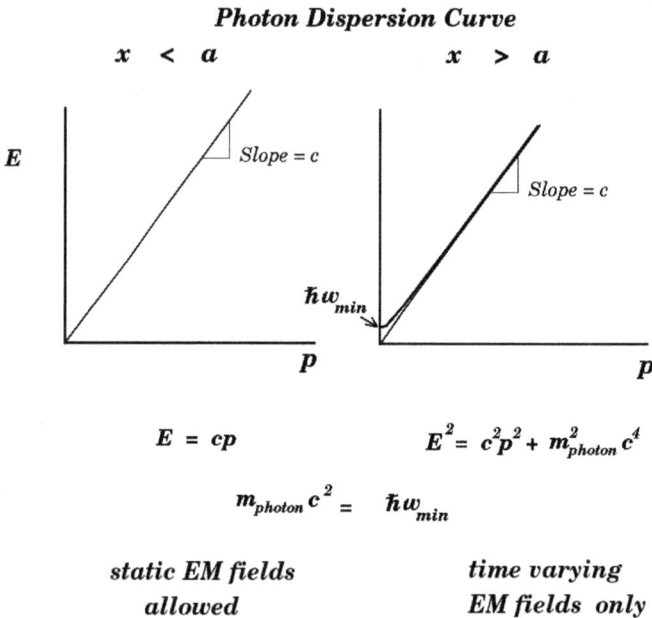

Photon Dispersion Curve

$$E = cp$$

$$E^2 = c^2p^2 + m_{photon}^2 c^4$$

$$m_{photon} c^2 = \hbar w_{min}$$

static EM fields allowed   **time varying EM fields only**

The left hand dispersion curve in figure 6.1 is measured on a scale that is very much smaller than the parameter a for the photon. At this scale the dispersion curve does not suffer the change of slope found in the relativistic mass energy relationship. Then at larger scales the self-energy is encapsulated in terms of mass, which is just a description of its mean interaction with vacuum and matter over an adequate volume. This results is shown on the right hand side of figure 6.1. The implications of this dispersion curve is that, the ability to have static fields on this large scale is lost. This is because the lower end of the frequency space is no longer accessible.

However, on a large scale, the net momentum of the minimum state is zero and it can become captive in a gravitational field. A photon interaction with the medium will be more active in high temperature plasmas where momentum can be transfered to the plasma to allow low energy transitions to occur. Thus, to radiate these low energy photons efficiently a dense high energy plasma may be required. These low energy photons are not easily produced nor are they active when in low density regions, except for their contribution to the local gravitational potential.

Summary

The photon is a particle we know the most about on a small scale. However, its not apparent that it is at all well understood on a large scale. We have provided a mechanism to show how bosons can acquire mass, and photons are bosons. The only way to gain experimental information on the photon mass at present is through astrophysical observation because the scales over which the photon can act on the vacuum state is vast if the lower bound photon mass estimates are used. If the photon would be shown to have a mass it would affect a number of calculations and deductions where the photon mass is assumed zero.

6.2.1   Photon Mass

In the recent NASA-Harvard-Smithsonian image[53] there is a high degree of symmetric perfection characteristic of a quantum mechanical stationary state, figure 6.2. This image is more characteristic of a quantum stationary state than an artifact of the activity surrounding plasma processes of a rotating massive black hole. Jets interacting with local plasma do not share this degree of bilateral symmetric perfection across 50,000 light years of space [54]. The photon on this distance scale can possess mass and then form a gravitational bound state with a black hole. The weak illumination of these bubbles in the x-ray and gamma ray region of the spectrum have more than one possible source and will be considered when the basic details of their structure is developed.

Two questions that need to be considered are how the particles become trapped and and what is trapped? The scale is assumed to be the gravitational potential from the galactic black hole. The mass of the boson in the bound state can be determined by the dimensions on the picture using a non-relativistic quantum mechanical description of a mass bound into a stationary state by the gravitational potential.

The problem that is solved is just the hydrogen atom where the replacement of the potential occurs:

$$-\frac{Ze^2}{r} \Rightarrow -\frac{GMm}{r} \tag{6.2}$$

Figure 6.2: **NASA-Harvard-Smithsonian image[53] "Energy Bubbles"**

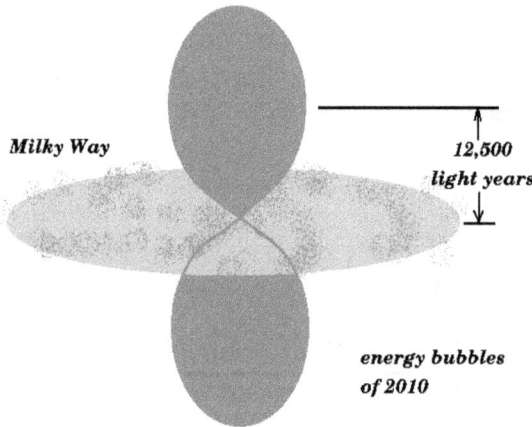

$$\widehat{H}\psi(r,t) = i\hbar\frac{\partial\psi(r,t)}{\partial t} \tag{6.3}$$

$$\widehat{H} = -\frac{\hbar^2}{2m}\nabla^2 + \frac{GMm}{r} \tag{6.4}$$

This problem is solved in detail in reference [1]. In the above potential for the black hole in our galaxy, M is $8.57 \times 10^{36} kg$ [55] and the small m is the individual mass of the particle in a particular stationary state. This problem is only interesting if the particles occupying the stationary states are bosons. In this case the candidate particle is the photon ($m << 10^{-11} \frac{eV}{c^2}$ or 10^{-52} kg)[56] because we are looking for something well below the mass range of the axion ($.001 \frac{eV}{c^2}$)[57] and the neutrino ($1 \frac{eV}{c^2}$) [41].

Where the wave function is $\psi(r,t) = u(r)e^{-\frac{iEt}{\hbar}}$ and in spherical coordinates this results in:

$$u(r, \theta, \phi) = R(r)Y(\theta, \phi) \tag{6.5}$$

Computing the relation for the energy eigenvalue is an involved derivation and can be found in Schiff [25] or Bethe and Salpeter [58] where the electrostatic terms are replaced by the gravitational potential. The resulting expression for the energy eigenvalues are as follows where $n = 1, 2, 3 \ldots$

$$E_n = -\frac{G^2 M^2 m^3}{2\hbar^2 n^2} \tag{6.6}$$

g_n is the scale factor that appears in the numerator of the exponential factors of the wave function that control the wave functions range can be defined:

$$g_n = \frac{\hbar^2 n}{GMm^2} \tag{6.7}$$

Table 8: Selected wave function components

n	$l = n$ -1	l -1
1	$e^{-\frac{r}{g_1}}$	-
2	$sin\theta e^{i\phi} \frac{r}{g_2} e^{-\frac{r}{g_2}}$	$cos\theta \ (1 - \frac{r}{g_2})e^{-\frac{r}{g_2}}$
3	$sin^2\theta e^{i2\phi} (\frac{r}{g_3})^2 e^{-\frac{r}{g_3}}$	$sin\theta cos\theta e^{i\phi} \frac{r}{g_3}(1 - \frac{r}{2g_3})e^{-\frac{r}{g_3}}$

The state of interest is the 2P state, where $l = 1$, because it has a zero amplitude at the black hole. All the S states would be consumed by the black hole because they have a finite amplitude at the origin. By finding the maximum in the radial density function $\rho(r) = r^2 R_{21}^*(r) R_{21}(r)$, which is the radially dependent density, for the 2P state determines the radius at which the density will peak. This peak in density is what is measurable from the bubble image.

$$\rho(r) \simeq r^4 e^{-\frac{2r}{g_2}} \tag{6.8}$$

Taking $\frac{d\rho(r)}{dr} = 0$, the maximum of the density is r_{max}

$$r_{max} = 2g_2 \tag{6.9}$$

and the mass of the bound particle can be calculated by setting $r_{max} = 12{,}500$ light years. The resulting mass of the particle in the 2P stationary state is 4.06×10^{-58} Kg. This mass is well below the experimental upper bounds from previous studies for the photon. The proof of whether these are bound photons or not will come from the analysis of the inverse Compton scattering [1] of these states to show that the particles scattered out of the states are photons. Since in this realm of low mass there are no other known particles that can exist in such quantities and be stable, it will be assumed that these states are occupied by minimum energy photons. Taking

$$mc^2 = \hbar\omega \quad this \ yields \quad \omega = 3.46 \times 10^{-7} \tag{6.10}$$

The cutoff frequency for the ground state photon is in a frequency domain that is difficult to study and is usually only reflected in fields on the dimensions of the Milky Way or greater.

---

[1]This appears to be a valid process yielding good estimates because the classical electronic cross section, Thompson cross section [59], is a constant at low energies. Also, with the photon having a rest mass, both the energy and momentum conservation allow the resultant interaction to produces a forward scattering of the photons. If the photon did not have a rest mass there would be no scattering.

In terms of computational importance, it provides a hard cutoff for perturbation calculations that integrate over all possible photon states. The self energy of $2.26 \times 10^{-22} eV$ for the photon is a minimum energy for any radiative transition.

The gravitational energy eigenvalues can be taken from these solutions and are:

$$E_n^{milky\ way} = \frac{6.124}{n^2} \times 10^{-33} eV \qquad (6.11)$$

These states, though they exist, have spacing that are well below the level of the minimum photon energy so that transitions between states are not possible. These states can only be accessed by transitions from the continuum. This is an important feature in that it makes the trapped photon a very weakly interacting captive particle.

A photon in a stationary state is the ultimate slow light. The photons being bosons can collect in enormous numbers in the stationary state and will be a form of dark matter that possibly can rival the mass of the galactic black hole. These variations in bound state masses may have much to do with the evolved morphological differences in galactic structure, such as the galactic bulge. Their density is spread over a large volume since it cannot be a point attractor. The stationary 2P state is imaged by radiative processes. However, it is apparent a weakly scattering bound state photon whose individual mass is 4.05×10^{-58}kg through an inverse photo electric effect or inverse Compton scattering is possible. The 1S state is not visible since it has a finite amplitude at the black hole and will be consumed. Also, it is possible that the mechanism for populating the states would not favor the 1S state. The 2P state has a zero amplitude at the black hole and, therefore, will not be easily captured. The black hole has all the characteristics of a singularity at that scale. Therefore, the S state photons, if they exist, have been consumed by the black hole. The S states are available for occupation, but that occupation time is limited by the black hole. The bubbles in the image are labeled as the P_z state where the P_x and P_y are in the galactic plane. The populating of a single oriented P state in a relatively mass free region may have more to do with the mechanism of angular momentum transfer from the black hole's mass acquiring process rather than affects limiting coupling transitions from the states that would lie in the galactic plane. The rate limiting process that determines whether these states are populated are the mechanisms for populating the P states not for transitions out of the state.

It is useful just to compare the scaling that gravitational forces can have on the size of a photon bound state. This is done for astronomical objects for which the masses are known. The volume of the mass has to be concentrated relative to the size of the stationary state to have any meaning so only compact objects are considered.

Table 9 Bound Gravitational Quantum State Scaling
g_1 *meters*

Central Attractor kg	photon 4.05×10^{-58}	10^7 photons 4.05×10^{-51}	axion trial mass 10^{-40}
Jupiter 1.9×10^{27}	10^{30}	10^{16}	10^{-5}
Sun 2×10^{30}	10^{27}	10^{13}	10^{-8}
SN1987a star 4×10^{31}	10^{26}	10^{12}	10^{-10}
Milky Way Black hole 8.57×10^{36}	10^{20} ($10^5 l.y.$)	10^6	10^{-15}
Massive Galactic Black hole 10^{42}	10^{14} ($.1 l.y.$)	1	10^{-21}

In addition to the photon, two other objects are considered. Particle or collections of particles with a mass of 10^7 times the photon mass and the axion with a proposed mass of 10^{-40} kg ($.001 \frac{eV}{c^2}$). Everything but the axion requires dimensions measured on the scale of meters and light years to gain a perspective of scale of the minimum stationary state. As the mass scales upwards, the dimensions required scale downwards by the inverse square of the mass so that the particle mass does not have to increase that much until the dimensions of the stationary states become modest.

The photon's mass has to be acquired from the perturbation of the vacuum. The perturbation of the photon that looks the most promising is one that produces a random phase delay over the interaction volume defined by the radius a.

Table 10 Boson Particle Scale for Acquiring Mass

particle	mass Kg	a meters	Cutoff Hz.
pAxion	1.8×10^{-39}	.1	-
Axion	1.8×10^{-39}	2×10^{-4}	6×10^{10}
photon	4.05×10^{-58}	$.89 \times 10^{15}$	5.5×10^{-8}

$$a = \frac{\hbar}{mc} \qquad (6.12)$$

The radius of the potential well required is on the order of 10^{15} m or .129 light years. That implies that mass measurement made in experiments on a smaller scale would result in a mass of zero being found for the photon. Since on a large scale

$$a << r_{max} \qquad (6.13)$$

the rest state photons will be well behaved gravitating masses in the much larger structure of the stationary state. The fact that an observable stationary state is available to a photon may say something about what actually composes the vacuum state. There is a long history of research on the photon and how it acquires mass reviewed in Jackson [15] with specific experiments and upper bound estimates [60], [61], [62],[63] and [64] . The experimental limits of measuring mass by the cited techniques are too great as compared to some estimates of the theoretical upper bound. A better understanding is required of the interaction of a photon with the vacuum state on a large scale to close this gap. Specifically determining a state function for mass of the photon, $m_{photon}(T, \rho)$ dependent on temperature and density found in stellar interiors.

Summary

Extreme displays of symmetry occur rarely on a large scale, so much could happen and long times are required to get things into place. Only a very strong driving force for a stationary state formation with a very stable weakly interacting captive could persist *for any length of time in the region around the center of our Milky Way. In order for such a state to exist, it needs a very light particle to be gravitationally bound into a quantum state that has zero amplitude at its center. Also, there has to be a mechanism to populate this state. The only evidence that photons could be the particles trapped in this state is that the radiation that illuminates this state like a weak veil should be proportional to the density of the state, and if the weak illumination is partially an inverse Compton scattering, then the trapped particles are photons. These photons have a computable mass of $4.05 \times 10^{-58} Kg$ measured from the volume they occupy.*

6.2.2   Galactic 2P State

There are two aspects of the 2P state's gravitational potential that are interesting. The first is why the limb of the 2P envelope is enhanced and the second is the effect on the gravitational potential at the galactic disk. A massive P state of photons that is rather uniformly distributed in what looks very much like a nearly spherical object will have an interesting effect on local matter that has a relatively low velocity compared to the galaxy. Matter gravitationally trapped by the stationary states' attraction will fall though the P state, gaining velocity, and then execute a simple harmonic motion through the P state in a collection of random orbits. The longest residence times for this trapped matter, τ_r, will be at the outer surface of the stationary states' mass distribution. The residence time is longest on the outer edge because the radial component of the velocity through the center of the P state goes to zero in that region before resuming its fall through the state. If the residence time is longest at the surface, the apparent density of matter be the greatest at the outer surface as well. The outer regions of the P state will be brighter because the increased density of non dark matter will scatter more radiation.

As the photon is a boson, and as the number of captive photons collect, the interaction terms for their gravitation attraction becomes significant both in its interaction with the galactic mass, and also with itself. This structure will be perturbed and relaxed by concentrating mass in the central regions of the $2P_z$ states above and below the galactic plane. This would result in a better defined apparent diameter of the bubbles that possibly can be measured. If the mass of the P states grew to be on the order of the mass of the central black hole, then there would be a significant change in the gravitational potential on the galactic plane. The potential on the plane would depend on an integral of the type

$$\phi_{2P}(R) \sim -2Gnm_p \int_0^\infty \frac{z^4 e^{-\frac{2z}{g_2}}}{\sqrt{R^2 + z^2}} dz \qquad (6.14)$$

Along the z-axis through the center of mass of the 2P state above the galactic plane the potential function become:

$$\phi_{2P}(Z) \sim -2Gnm_p \int_0^\infty \frac{z^4 e^{-\frac{2z}{g_2}}}{\sqrt{(Z-z)^2 + \frac{Z^2}{4}}} dz \qquad (6.15)$$

where R is the distance out from the center of the galactic plane, n is the number of photons in the state and m_p is the photon rest mass. Also, g_2 is on the order of R when R is at the galactic periphery. This integral will produce a much smaller decay of the gravitational potential as a function of distance from the center. A comparison plot is shown below for the roll off in the 2P gravitational potential verses the $\frac{1}{r}$ potential of the galactic black hole.

For the P state to significantly affect the gravitational potential on the galactic periphery, accelerating stellar orbital velocities, the total mass of the P states should be on the order of the black hole mass or greater. This potential would contribute to increased radial velocities at the periphery of the galaxy to help explain the deductions of F. Zwicky, [65] [66], who originally proposed dark matter to solve the problem of a gravitational potential that did not fall off so rapidly. Actually, in order to fulfill F. Zwicky's original goal of accelerating the motion of the periphery of the galaxy, dark matter alone would not be enough. That matter

Figure 6.3: *2P state gravitational potential is summed with the black hole potential and plotted in the plane of the galaxy. The 2P state has the same mass as the black hole and as a reference the central black hole potential. The visible radius of the Milky Way is the maximum range in the plot.*

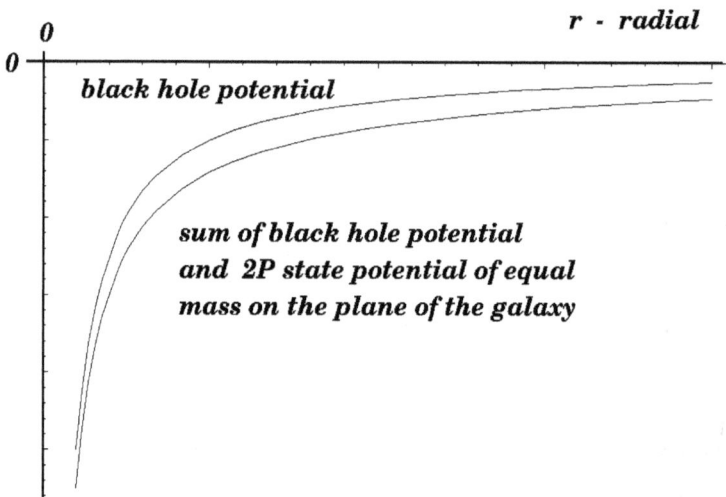

would have to be a stable structure over time in order to satisfy the dynamical requirements deduced from observation. This means a quantum stationary state might be the only structure to supply that stability. From the velocity distribution of stars in a galaxy it should be possible to calculate the mass of the galactic 2P state directly. A galaxy with mass distributed in multiple P states, 2P, 3P and 4P, the potential curves would flatten for higher N values and yield a stepped structure in the rotation rate as a function of radius [67]. There should be nothing special about our galaxy and these should be common structures, but in galactic collisions, the mechanics and outcome of these processes should be strongly affected by the presence of the P states. Evidence of the P states should be found in the detritus of these large atomic like collisions.

Figure 6.4: *2P state gravitational potential is summed with the black hole potential and plotted on the z axis above the galactic plane. The 2P state has the same mass as the black hole and the central black hole potential is plotted as a reference. The visible radius of the Milky Way is the maximum range of the plot.*

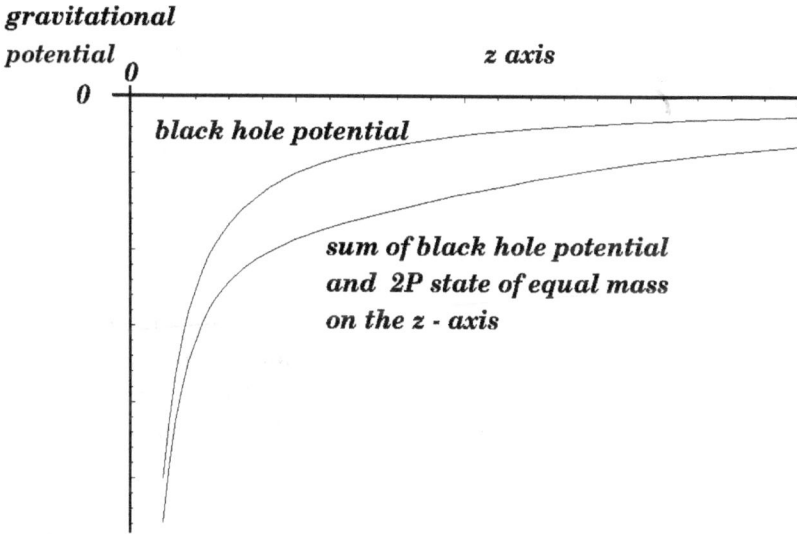

Using the potential functions it is possible to sketch the force on a test mass of a rather massive P state in figure 6.5.

At first glance this shape is similar to some of the halo structures found in the catalog of galactic images[68]. Because of the diffuse nature of the mass in the P-state, its gravitation potential decreases closer to the center of density until it vanishes for a particular lobe. This loss of a strong central attractor for the P-state can result in this diffuse bulge.

The forces on a mass captured by the gravity of the black hole and the P-state will be biaxial. If the P-state were not so massive then the gray region would encompass the state itself. Within this gray region trapped gas and dust will orbit in a complex pattern being affected by the attraction of both the central black hole and the stationary state. If the P state is heavily populated

Figure 6.5: *The zero Z direction force component for a star and 2P state of equal mass defines the envelope of the gray regions.*

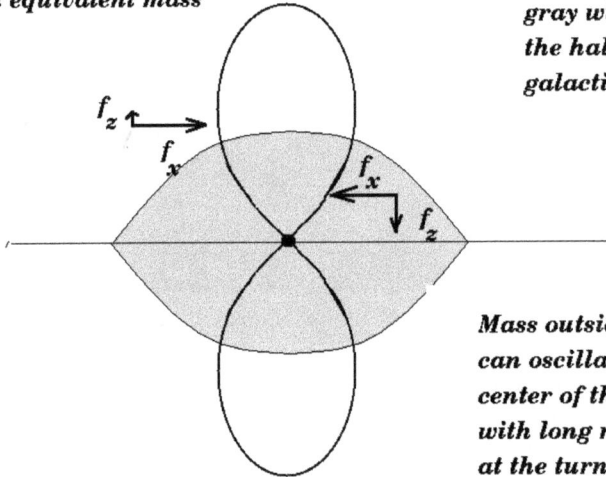

P-state and Black Hole with equivalent mass

f_z

f_x

f_x

f_z

Any mass within the gray will orbit in the halo of the galactic center

Mass outside of the gray can oscillate through the center of the P-state with long residence time at the turning points.

and a strong attractor, then the surface of axial force, $F_z = 0$, which makes up the gray envelope, will define the orbits of matter trapped by the joint potentials. Because constraints on the potential the familiar precession of a classical top would begin to operate [29]. This would place matter in an orbital band around the P-state even if the P-state's mass was substantially less than that of the black hole. This collection of matter could then be imaged as a ring under the proper conditions.

6.2.3   Bound State Spectroscopy

The question of what mass an attractor must posses to form a bound state of photons is one way of approaching the problem of how these states can be populated. There appears to be no minimum mass threshold, though there is a maximum mass above which would encroach upon the minimum volume requirement for the photon to acquire mass. For a small mass, this state would be enormous. The question then becomes how could you populate such a state? Populating the states becomes a problem since the transition level spacings are well below the threshold of a minimum photon energy even for the Milky Ways galactic black hole so if the states are populated, there can be no simple transitions between states. Populating the state requires a process where the system loses energy that results in a very low energy photon created within the gravitational field that remains bound. This can result from processes driven by a gravitational potential from figure 4.4 where minimum energy photons are emitted. This results in a populated state which is an extremely inert object. There can be no direct radiative transitions between states except possibly by a Raman like process. The large size of the photon states and their very weak magnetic and electric field, which are averaged over these large volumes, reduces the coupling to local matter except from inverse Compton scattering.

Figure 6.6: *Available states of gravitationally bound photons to a black hole*

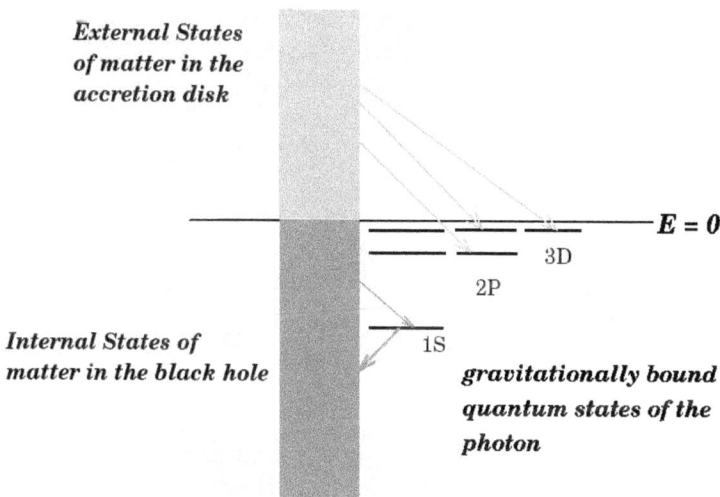

External States
of matter in the
accretion disk

$E = 0$

3D

2P

Internal States of
matter in the black hole

1S

gravitationally bound
quantum states of the
photon

The one feature of the 2P state is that it provides a reservoir for angular momentum when dense gravitating objects spin down to allow matter to move to their cores. The bound state provides a sink for angular momentum from radiating massive particles and passing photons.

In the strong gravitational field in the region of the black hole charged particles and electromagnetic radiation will produce low frequency radiation, removing mechanical angular momentum to supply the black hole states with bound photons. This process requires energy being radiated into these minimum energy photon states, and the best source of that radiation is from the hot and high density plasma that will be opaque to all radiation except these minimum energy photons, which can bleed energy from this astrophysical compressor.

Figure 6.7: *Mass accretion driven radiation which reduces mechanical angular momentum and energy captured from passing photons which populate a gravitationally bound P State*

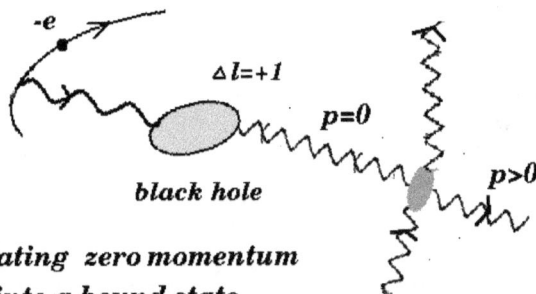

Mass radiating zero momentum photons into a bound state shedding angular momentum in a strong gravitational potential.

Bending light

The coupling to these extended photon, minimum energy states in vacuum will be very weak. However, within the dense plasma of very hot stars central regions or within an accretion disk where the plasma electrical conductivity is high these states will be much more compact and in numbers, if coherent, will strongly interact with the plasma. A simple scattering model will probably not be sufficient to solve the spectral transition problems that are of interest.

Summary

One form of dark matter, the collection of large quantities of photons gravitationally bound in a quantum stationary state, is a form of matter that must be considered. The main difficulty with this form of matter, if the photon has a mass, is populating the available stationary states. Since the numbers of photons required to make an impact on the amount of dark matter is on the order of 10^{93} photons the process that produces them must be highly efficient. This is a great deal of energy and matter and it must come from someplace. One of those places is the mass accretion process for a black hole, where both energy and angular momentum can be extracted.

6.3   pAxion

For bosons we have two examples, the photon, with an intrinsic angular momentum of one, and the pAxion, with a mass about 10^{-9} that of an electron. These two objects represent the lowest mass objects to test the concepts developed in the previous chapters. These are different objects and we expect the details of their potential interactions to be different.

The pAxion has a propagating state with a parabolic dispersion curve of a classical particle with mass. However, in the case of the pAxion this mass is probably a result of a weak interaction with the more massive spin wave population. With increased population density many pAxions can collect into a bound state that is a Bose-Einstein condensation. These two distinct behaviors are seen in the reduced dispersion curve data [5]. This complex behavior is more apparent than that of the photon because of its relatively stronger interaction with the medium. A coherent collection of pAxions can form a bound state with just the potentials they generate interacting with the ferromagnetic medium. This effect can encompass a large volume. This produces an interesting array of properties characteristic of their quantum origin on a large scale. The pAxion's scattering processes are very different than what can be expected from a photon because of the stronger interacting medium. These weak field measurements in a magnetic material are easy to perform and we have recently noticed similar programs being extended to photons measured in transmission [69].

The third example that will briefly be considered is the excitation that carries current in graphene, which like the photon, has a dispersion curve of a massless particle. The question for the carrier in graphene becomes whether the sample volume, over which measurements are made, is too small for an excitation to develop an effective mass. Or whether the lattice bondings are so strong that there is no available excitation to affect the particle. On the micron scale where conduction is studied for graphene the electron and hole carriers may not have the space to separately generate the statistics of either group of particles and just remain a simple coupled excitation. Both the photon and the carriers in graphene have a relatively high finite velocity and that in itself suggests the mediums are relatively non compliant and any process that generates mass is a low probability event on a small scale. The interesting feature about graphene is that it is a two dimensional system with an effective mass of zero. The loss of a dimension will decrease the ability of the lattice to support the excitations necessary to generate mass on a small scale. In the case of graphene the dimension is determined by the carbon lattice. In free space the only way to get a two dimensional interaction is to have two free particles interacting via a central force which are naturally constrained to interact within a plane unless they form a bound state.

Summary

The pAxion being an excitation of iron based alloys is interesting to study because it is weakly interacting much like light in a vacuum. However, it is bound to a cubic crystal system that has complex magnetic domain structures. It can be easily studied and has features that can test some of the model arguments that have been introduced. The photon and graphene conduction carriers have more in common on a small scale, being massless, than they do with the pAxion which is a slightly stronger interacting species.

6.3.1   pAxion Characteristics

The pAxion measurement found three separate states, two propagating states and a bound state [5]. It is assumed that the bound state generated its own potential well with a population of exciton forming an attraction and it has angular momentum zero with the maximum of its wave function at the field's source. From a primitive model which generates a linear oscillating magnetic polarization, it was assumed that the low energy bound state had an angular momentum of zero. So that a zero angular momentum state with a wave function $\frac{sin(kx)}{x}$ where $k = \sqrt{\frac{2M(V-|E|)}{\hbar^2}}$, the size of the state will scale as $\frac{1}{k}$ and as $E = \hbar\omega$ shrinks so does k. For a free an unbounded wave function, the energy dependence is reversed and as E increases the wave function shrinks; therefore, it is possible to launch a high frequency component and scatter it off a low frequency target detecting the by products simultaneously. The experimental details of the two specific scattering measurements are in reference [5] and Appendix C. The question becomes whether higher angular momentum states are associated with the propagating modes? One way to investigate this is to do a scattering experiment with two signal sources. The target strength can be changed by the amount of current injected into the target coil.

In observing these propagating fields, they appear to be unaffected if a region of the bar in which they are moving is magnetically saturated with a strong static field. Magnetic saturation effectively reduces the number of domains to a few large ones that are on the scale of the material sample size. If, however, a region of the bar is raised above the Curie point, the bound state exponential spread is quenched and only the propagating component is passed. The propagating component, once launched, is little affected either by saturation or having the material above the Curie point. Also, the fact that the bound state component did not go beyond the region held above the Curie point indicates this was dependent on the magnetic medium which allowed the bound state to be formed. This bound state is assumed to be a Bose-Einstein condensation.

The minimum frequency for radiation based on the self energy is defined as $mc^2 = \hbar\omega$. This characteristic frequency is sometimes referred to as an infrared cut off [52]. For the pAxion, a cutoff would be difficult to observe. As the velocity of the pAxion falls below the speed of sound it becomes coupled to elastic waves, which have no cutoff. More importantly, the speed of light in a magnetic material varies as $c^2 \sim \omega$ which implies there is no low frequency cutoff for the pAxion so that it should be observed to zero frequency. This is confirmed by many measurements. The zero frequency limit is a static ferromagnetic ordered state that does exist in iron and steel.

Summary

The pAxion appears to be a boson which has multiple forms. One of which is strongly dependent on the net magnetization of the medium and one component that is independent of that magnetization. That implies the attractive interaction between components to form a BEC is lost in the high temperature region.

6.3.2 pAxion in a Wire

The simplest experiment performed on a 12 mm diameter hot rolled low carbon steel wire was to measure the propagation properties verses frequency and as a function of detector displacement

from the field source. One of the interesting pieces of data from
this dispersion curve experiment appeared at the upper end of
the measurement range at 2 and 3 MHz. In these measurements,
there is a fixed source coil and a measurement coil that is trans-
lated down a low carbon steel wire. These measurements are fully
described in [5] along with the data that will be used in this section. The data away from the the source coil decays very slowly
and shows two unexpected features. First, instead of a monotonic
decreasing field measurement with distance, the response actually
shows an increase after decaying over the first quarter of the sam-
pling length for a distance and then flattens out as shown in the
amplitude response in figures 6.8 and 6.9. In the phase data, there
appeared a modulation pattern with a high frequency component
and a low frequency component. The first figure 6.8 is a plot of
the log amplitude data verses the sensor position from the source
field, which looks very well behaved. The next figure 6.9 shows a
linear plot of the same data enlarged to show the deviation from
monotonic behavior. This feature is best seen in the 3 MHz data
where there is actually an increase in the response. If these are coherent linear magnetization excitations, then it appears that one
component in the long range field has $l = 1$ or greater, these wave
functions start at zero at the origin and grows into an oscillation
as presented in Table 1. It is a small contribution in the data, but
well above the noise floor of the measurement system. The linear
geometry of the wire limits the accuracy by using the full three
dimensional wave function.

The more definitive data is the phase data. The first graph presented in figure 6.10 covers the entire span. Then in an expanded
form showing the oscillation on a scale well above the system noise
in figure 6.11. This data is of particular interest because the detected phase variation is very difficult and almost impossible to
induce by mechanical means. A phase change of this type would
typically be generated by a material variation in either conductivity or permeability as a function of depth into the material.
However, in a material like a mild steel, these variations if they
exist are usually seen at much lower frequencies of a few kilo hertz

Figure 6.8: *Raw amplitude dispersion data for the pAxion scaled normally, source at x= 0*

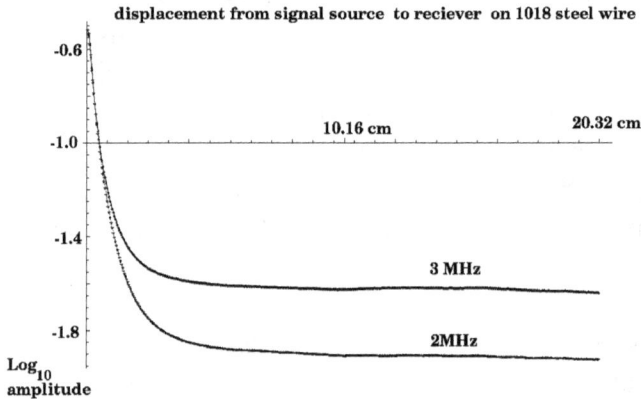

Figure 6.9: *Raw amplitude dispersion data for the pAxion scale expanded to show amplitude decay is not monotonic*

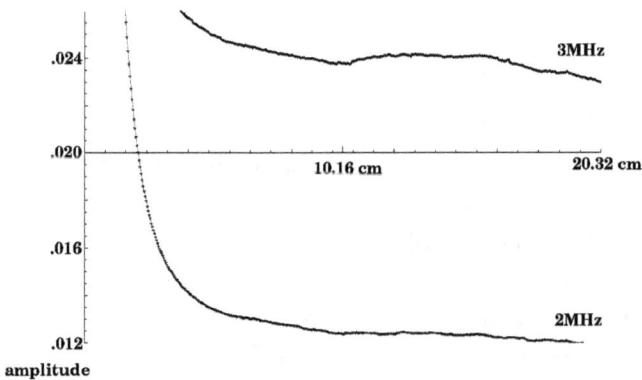

Figure 6.10: *Raw phase dispersion data for the pAxion scaled normally to show phase delay is not monotonic*

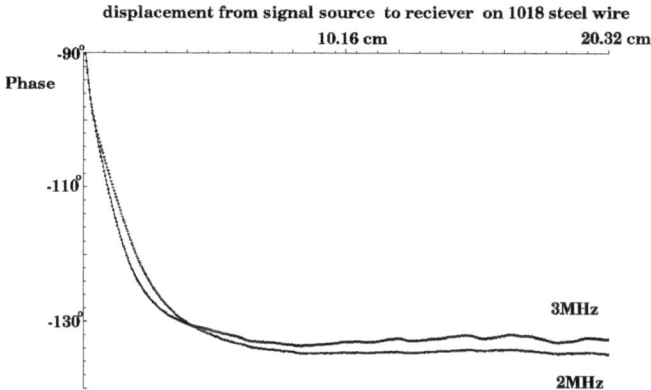

and not detected at frequencies above 50 KHz . So the detected oscillation in the phase data as a function of linear position appearing almost frequency independent was very unexpected. This coupled with an amplitude change that is not monotonic in a homogeneous material suggested an interference process at the level of the wave function. The data implies that there are multiple fields active that can produce a changing phase response that is not monotonic.

The one case we would expect to detect a phase oscillation pattern requires two conditions to be met. The first condition is the phase of at least two wave functions that produce the density function are phase shifted from each other. This could occur if the sourcing of the $l = 1$ state were displaced from the $l = 0$ in the source volume. The other requirement is that the propagation vector for the two angular momentum states be only slightly

---

[1]The inspection induction responses of steels is primarily in the geometrical variations of surface structures such as seams and cracks which can be performed over a similar frequency range. This allows detection of features in the micron range without a significant contribution from material property variations. At these higher frequencies the material variations are compressed into the asymptotic high frequency pole and phase variations are suppressed at the $-180°$ asymptote [5].

Figure 6.11: **Raw phase dispersion data for the pAxion scale expanded to show phase oscillations**

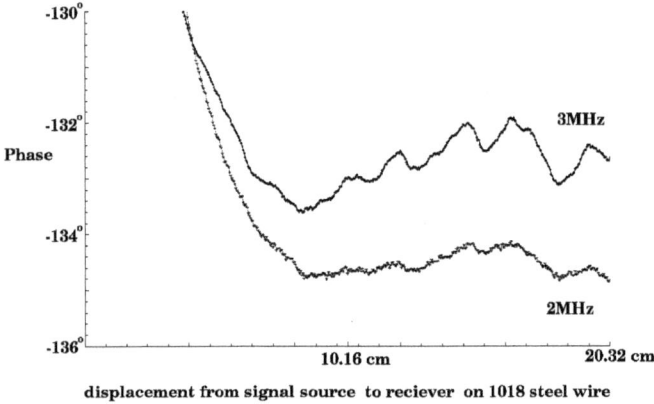

different. If there were no phase difference between the wave functions, there would be no interference pattern. The propagation vector k_l and the phase ϕ_l are the key parameters for the wave functions $\psi_l(x)$ in the bound state. Using the spherically symmetric solution one immediately has the $l = 1$ and $l = 2$ solutions, which are zero at the source and growing where as the $l = 0$ has its maximum at the source and decays as a function of length. This behavior is mirrored in the data if at least two components are present. The smaller variation in the 2 MHz data looks to be due just to a reduced fraction of the higher angular momentum states. The data actually looks to be composed of three components, at least, because the high frequency periodicity is suppressed in some regions. If the $l = 2$ state were also present, it could be used in modeling this wave form. The variation in the propagation vector between the three components could arise from the different way each component couples to the wire. This difference appears to be small because of the measurable variations in the phase data. The phase variation between components would be determined by the mechanism that spawned each of the components. The three components can be modeled in this wave function that is the sum of the three components.

$$\Psi = a_o\psi(k_ox)e^{i\phi_o} + a_1\psi(k_1x)e^{i\phi_1} + a_2\psi(k_2x)e^{i\phi_2} \qquad (6.16)$$

The terms break up into two groups when computing the density.

$$\Psi^*\Psi = \Psi_o^*\Psi_o + \delta\Psi^*\delta\Psi \qquad (6.17)$$

The main measured data vector is determined by the product $\Psi_o^*\Psi_o$.

$$\Psi_o^*\Psi_o = \sum_{l=0}^{2} a_l^*a_l\psi(k_lx)^*\psi(k_lx) \qquad (6.18)$$

The correction to this due to variations in the propagation vector and the relative phases at which the fields are generated are contained in the terms.

$$\delta\Psi^*\delta\Psi = a_l^*a_j\psi(k_lx)^*\psi(k_jx)e^{i(\phi_j-\phi_l)}$$

$$(6.19)$$

$$for\ all\ l \neq j;\ l,j = 0,1\&2$$

This relation produces six terms. The coefficients a_l are real. The complex character is included in the $e^{i\phi_l}$ factor. We have not solved this problem for the specific case of a wire with a finite radius, but in order to get an approximate result on the phase behavior we will use the three-dimensional solutions projected on the wire. Using the first three angular momentum solutions from Table I while keeping only the larger terms, those that fall off the slowest as $\frac{1}{r^2}$:

$$\delta\Psi^*\delta\Psi\ for\ \frac{1}{r^2}\ terms =$$

$$-a_oa_1\frac{sin(k_or)cos(k_1r)}{r^2}(e^{i(\phi_1-\phi_o)} + e^{i(\phi_o-\phi_1)})-$$

$$a_0a_2\frac{sin(k_or)sin(k_2r)}{r^2}(e^{i(\phi_2-\phi_0)} + e^{i(\phi_o-\phi_2)})+$$

$$a_1a_2\frac{cos(k_1r)sin(k_2r)}{r^2}(e^{i(\phi_2-\phi_1)} + e^{i(\phi_1-\phi_2)})$$

$$(6.20)$$

The first four terms reference to the vector of the main response at ϕ_o, and they can only produce an amplitude variation by the factor of $cos(\phi_i - \phi_o)$. However, the last two terms have vectors that are not bound to the ϕ_o phase and will produce two vector responses whose phases differ from that of the main response. It is these two terms that will produce the phase oscillations in the measured phase response of the dispersion curve. For this phase variation with distance along the wire to occur both the $l = 1$ and $l = 2$ angular momentum states must exist.

Summary

This is almost the simplest experiment one can do by constraining a single frequency propagating field in a wire, 12.7 mm diameter and observing its decaying behavior by moving the detector away from the source. The results here stood out because the phase oscillation is a phenomenon that neither common or easy to induce under any circumstances. For the phase to change monotonically all that is required is a delay due to signal propagation, but to have an oscillation in multiple components not coupled to the main field are required. It appears there are at least three signals present and this does not disagree with our initial finding that there may be three stable angular momentum states that this boson can assume.

6.3.3   pAxion Scattering in a Plate

In order to investigate whether there are transverse components of the pAxion field it is difficult to do such experiments in a wire. However, it is possible to perform the measurements in a plate. Also, the constraints of the wire will most probably force a change from a three-dimensional field response into a response that takes on a one dimensional form for the propagating field directions. A three dimensional solid presents some challenges to work in and would have to be assembled from components and would not be monolithic. The plate experiments are quite simple and use the launch and detection techniques previously described for wires to produce and measure the fields.

Figure 6.12: *pAxion transverse scattering experiment in a low carbon steel plate*

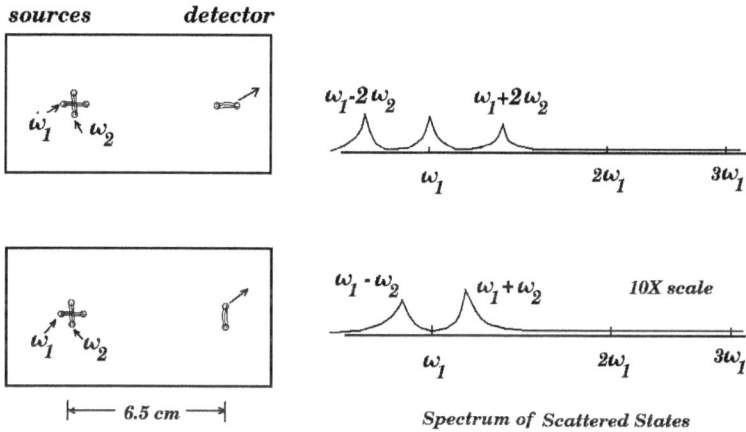

Spectrum of Scattered States

This is an experiment that scatters one field from another with orthogonal polarizations. One purpose of the experiment was to identify if there exists an orthogonal polarization for the pAxion. The inductors were made with no. 26 enameled copper magnet wire wound through two 1.5mm holes that were drilled on 1 cm centers and threaded with 5 turns. The experimental set up is shown in figure 6.12 where ω_1 is 49 KHz along with the responses due to the applications of the second source ω_2 at 2.5 KHz. The results only reflect changes in the detected signal when the second source was injected, and the raw data is presented in the tables that follow the figure. The data is presented as a two dimensional vector measuring a voltage (V_x, V_y), where the x-component is in phase with the source oscillator and the y-component is -90° delayed from the source oscillator.

Table 6: pAxion Axial Scattering in a Plate, (0° Comp., 90^0 Comp.) in relative units, first as (x,y) components and then in vector representation. Note that the ω_2 field is phase locked to the probe field ω_1. The raw quadrature data are located in the first two rows in relative units. $\omega_1 = 49KHz$ and $\omega_2 = 2.5KHz$

test	$\omega_1 - 2\omega_2$	$\omega_1 - \omega_2$	ω_1	$\omega_1 + \omega_2$	$\omega_1 + 2\omega_2$
no ω_2	-320, 128	-320,120	-100590, 26000	-350,110	-330,150
with ω_2	1165, -1458	-327,39	-100268, 27835	-290,110	635, -758
difference	**1485,-1586**	-7,-8	**322, 1835**	60,0	**965,-908**
vector original	-	-	100%, 165.51°	-	-
vector result	2.1% −46.88°	-	1.8%, 80.05°	-	1.3%, −43..26°

Table 7: pAxion Perpendicular Scattering in a Plate, (0° Comp., 90^0 Comp.) in relative units, first as (x,y) components and then in vector representation. Note that the ω_2 field is phase locked to the probe field ω_1. The raw quadrature data are located in the first two rows in relative units. $\omega_1 = 49KHz$ and $\omega_2 = 2.5KHz$

test	$\omega_1 - 2\omega_2$	$\omega_1 - \omega_2$	ω_1	$\omega_1 + \omega_2$	$\omega_1 + 2\omega_2$
no ω_2	-325, 65	-330,65	7665,11165	-355,65	-335,80
with ω_2	-360, 105	-470,245	7650,11135	-575,190	-365,105
difference	-35,-40	**-140,-180**	-15,-30	**-220,125**	-30,25
vector original	-	-	100%, 55.53°	-	-
vector result	-	1.7%,−127.86°	-	1.9%, 150.40°	-

When the probe field, ω_1, and the detected field share the same polarization the single frequency interactions ($\omega_1 \pm \omega_2$) are essentially zero. The single frequency scattering satisfies both energy and momentum conservation if

$$\mathbf{k}_1 \bullet \mathbf{k}_2 = 0 \qquad (6.21)$$

which is a statement that the fields be perpendicular. This is the case for the geometry of the fields in the plate. In the case of the double transitions the conservation condition is more complex, where the two ω_2 pAxions are labeled a and b.

$$\mathbf{k}_1 \bullet \mathbf{k}\,_2^{\,a} + \mathbf{k}_1 \bullet \mathbf{k}_2^b + \mathbf{k}_2^a \bullet \mathbf{k}\,_2^{\,b} = 0 \qquad (6.22)$$

The simplest case for this to occur is if the three fields have propagation vectors that are mutually perpendicular, but there is probably not a mechanism to produce that result. This would result in a field ($\omega_1 \pm 2\omega_2$) that would have a component not aligned with the probe field's axis, ω_1, and could only be detected in the transverse detector. This does not occur, so that particular mode is not active in this thin plate. The symmetric emission or absorption as shown in figure 6.13 is an allowed transition that preserves the axial orientation of the output beam and maintains conservation of momentum and energy.

Figure 6.13: *Two pAxion induced emission process*

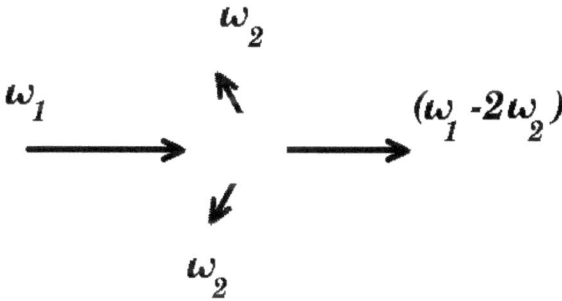

Secondly, the probe field's amplitude has increased and is altered by the addition of approximately 1.8% in magnitude with a phase that is shifted 85° from the original field. The phase of this additional gain is close to the transverse amplitude measured from the probe field. Because of the magnetic field's nature in a ferromagnet, one should expect both an axial and a transverse polarization field to be produced. When looking in the transverse detector, if no transverse field was produced, then the measured amplitude in that detector would be expected to be less than one percent of the axially polarized field. This is not the case for the measured amplitude, which is 10% of the axially polarized field and has a phase shift of 115° shift from the axial field. The process producing this gain in ω_1 is most likely a two step process, which will satisfy the first momentum conservation condition.

$$\omega_1^{\perp} + \omega_2^{\|} \rightarrow (\omega_1 + \omega_2) \rightarrow \omega_1^{\|} + \omega_2^{\perp} \qquad (6.23)$$

and

$$\omega_1^\perp \to (\omega_1 - \omega_2) + \omega_2 \to \omega_1^\parallel \qquad (6.24)$$

These two processes can move a small population from the $\perp$ field to the $\parallel$ field. The fact that the phase of this contribution is close to that of the perpendicular field indicates that the original source for those additional pAxions came from the perpendicular field. A fraction of that field transformed into a parallel component results in a gain for that channel measurement.

A good way to visualize this data is to present it in a polar plot, figure 6.14, is a polar log scale vector representation where all the responses are compared to the axially polarized probe field $\omega_1^\parallel$ detected with a sensor that is oriented parallel with respect to the source. Because the probe field and the scattering fields are derived from the same master oscillator, they are phase locked and it is possible to compare the phases of the resultant fields. The phase of the perpendicular component of the probe field is slightly more than 90° out of phase from the larger parallel component of the field.

Figure 6.14: *pAxion transverse scattering summary showing phase relationship between components*

Scattering Response Referenced to $w_1\parallel$

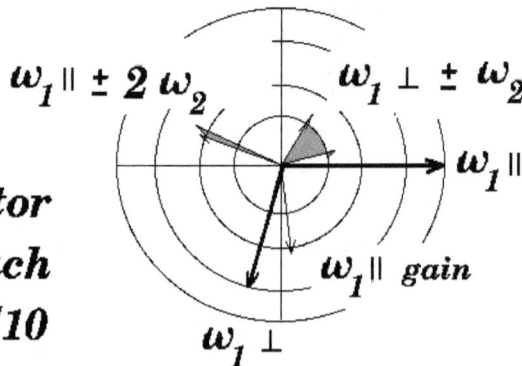

$w_1\parallel \pm 2\, w_2$ $w_1 \perp \pm w_2$

$w_1\parallel$

Log Vector
scale each
circle 1/10

$w_1\parallel$ gain

$w_1 \perp$

The fact that the scatter states are almost 180° out of phase from the source fields is characteristic of an electromagnetic reflection. In this case there is no reflection, just an absorption or emission of a much lower energy excitation. This is probably due to strong interaction with the more intense low frequency fields that are driving the formation of a local BEC at ω_2. In order to see these fields, the current injected into the ω_2 source was increased until the single frequency transitions were detected.

An important feature in plate measurements is the existence of the transverse component of the B field and the measured phase shifts of the scattered fields. The phase variations between the incoming and scattered waves in the plate experiment are large.

Summary

The pAxion scattering interactions are interesting because they can be used to probe the nature of the pAxion polarization modes. Two distinct scattering modes can be extracted and these indicate that two polarization modes for the field exist relative to the axis of propagation. The phase change from the scattered state to the initial state is not completely understood and should contain information about the details of the coupling to the lattice. We have not commented on the specific interactions that generates mass for the pAxion but its coupling to weak bulk currents are the most likely the origin of its mass value.

6.4   Supernova 1987a

There is one other object with characteristics that indicate a large scale stationary state may be populated. However, it is more complex than the bubbles trapped by the Milky Way because the gravitational source is not as massive as a galactic black hole and its stationary states overlaps the star which is the attractor. Testing the idea of a gravitationally induced stationary state on sn1987a surrounding structure seemed natural because of the observed rings. A single equatorial ring of gas may not be considered to be unusual except for its scale; however, a family with two rings displaced from the equatorial plane eliminates a simple central mass gravitational model for their origin. In fact, what we find unusual is the equatorial ring of gas at distance where matter is not easily concentrated for a star. If the pair of more distant rings is a feature produced by a P state, that bound state would simply be a sink for the angular momentum shed by the star to allow a collapse when the thermal energy was depleted.

The supernova 1987a contains four features and characteristics that are of interest. It has two families of rings, with a good estimate of the mass of the original star and the rings have an axial symmetry. The inner ring which is now brightening as matter from the supernova is colliding with that gas and dust and is not a simple shell but a ring. For a 1S state with a P state the expectation would be to find a uniform shell of gas and dust that would concentrate on the periphery of the lowest possible bound state. On the first look, the ring structure is sufficiently similar to where regular matter would be expected to collect, being attracted to a stationary state structure of a 1S and single 2P state on the planes of $F_z = 0$. Where F_z is the gravitationally determined force in the z direction along the axis of the rings. With the accurate dimension of the ring structure [70] being available along with an understanding of the thermal state of the star [71] that was the source of the supernova, a more detailed analysis can be made of this structure.

Supernova 1987a [72] has an extended viewing history. This supernova may supply some important data on the allowed states of dark matter because normal matter that has collected was illuminated. Figure 6.15 is a redrawn version of an image of supernova 1987a taken in 1994. In the photograph there are 3 resolved rings and the central object. The striking feature is their perfection in the extreme circumstance of being illuminated by a supernova. Because of the large scale perfection, it is felt that these are structures due to the effects of a quantum stationary state. However, from the point of view of a bound low mass particle where gravity has collected mass into the visible rings, this is a much more complicated object because the the quantum bound states overlap a large and hot stellar plasma core. The reason this is a complication is that the photon mass that was derived from the image of the energy bubbles of the Milky Way was that of a 2P state and that state occupied a large volume of relatively empty space. The case for the supernova is not the same, because the star in its final phases is quite large and hot with a significant overlap of any bound state.

Also there is the feature of the paired outer rings, that implies there was an axial symmetry associated with the star. This most probably was the star's angular momentum axis. If there was angular momentum transfer by radiation into a P state, the axial symmetry could have been transfered from the mechanical angular momentum of the star to the bound state.

Figure 6.15: *Supernova 1987a with its rings that define the gravitational potential fields*

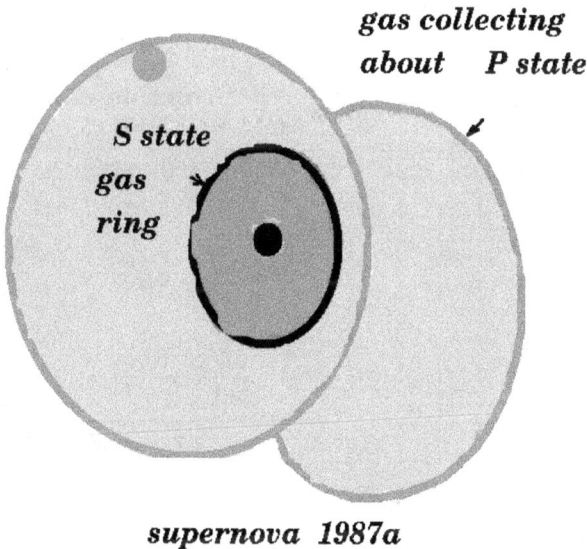

supernova 1987a

The star that gave rise to the supernova is lacking the black hole found at the center of the Milky Way. This allows occupation of the S stationary states, which will have a finite amplitude in the center of the stars. The three ring pattern radius ratios are not close to the 1S and 2P means size ratio $\frac{10}{3}$. The inner single ring has a radius of .33 light years and the radius of the outer pair is .73 light years [70]. If the 2P rings represent the location the $F_z = 0$ surface from gravitational attraction, then it would be expected to have a reduced ratio as compared to the mean radius. This is because that surface is closer to the star because of gravitational null surface is less than the $\frac{10}{3}$ ratio. With the radii and the mass of the star it is possible to compute the mass of the particle trapped in the gravitational potential for all the candidate states and this is done in table 11.

The ring's of supernova 1987a and the energy bubble of the Milky Way are on two very different scales, but most likely both reflect the interaction of normal matter with the frozen out stationary states of gravitationally bound very light particles. In the case of the Milky Way, the candidate particle is the minimum energy photon. Those photons require a large volume of vacuum to occupy.

Table 11 Computed masses for light particles, where the underlined items are the most probable candidates.

Trial State	measured size for g_n in light years	computed mass Kg.	Size Ratio
<u>1S</u>	.33	<u>3.66×10^{-51}</u>	1
2S	.33	5.17×10^{-51}	4
3S	.33	6.34×10^{-51}	9
<u>2P</u>	.73	<u>3.47×10^{-51}</u>	<u>3.33</u>
3P	.73	4.2×10^{-51}	8.33

The masses of the trapped particles in the 1S and 2P states are about 10^7 times the rest mass of the photon computed for the Milky Way 2P state. This can be due to four things: first, there is a new particle with this mass that occupies the states, second, there could be a collection of BEC made of 10^7 photons each,

third, the overlap of the wave function with the star is sufficient
to generate the added mass of each photon or finally none of the
above. The last case is improbable because the overlap of the 1S
and 2P functions with the stellar core are very different. For low
energy photons we don't have a mechanism to generate a potential
to form a BEC making the second case doubtful.

Table 12 Source of sn1987a Structure

Source	Comments
stellar core enhanced photon mass	1S and 2P core overlap is different
BEC of 10^7 standard photons	no known mechanism as yet
new massive weakly interacting particle	$l = 0$ boson possibly

The new particle would have to be boson and it is well in
the mass range where axion searches have eliminated a low mass
particle. But from our solutions for allowed boson states, the $l = 0$
can be much more compact than the $l = 1$ for the standard photon.
This should allow the $l = 0$ hybrid to have a much greater mass.
At this mass the size parameter, a, would be 10^8 meters which is
smaller than stellar core and it would behave as a massive particle
in a 1S state with a dimension of .33 light years.

These more massive minimum energy photons would more easily quench the thermal energy available in the core of the star. The
reason for this is the more massive photon would have a much
higher cutoff frequency by a factor of 10^7. That would increase
the radiation rate into the state by a factor of 10^{14}. The radiation
rate for $\omega \rightarrow \omega_{min}$ as a function of frequency and temperature is:

$$I(\omega, T) = \frac{\hbar\omega^3}{4\pi^2 c^2 (e^{\frac{\hbar\omega}{kT}} - 1)} = \frac{\omega^2 kT}{4\pi^2 c^2} \tag{6.25}$$

In order for a gravitationally bound stationary state to exist,
the volume of the final state must be greater than what is required
by the particle to acquire mass. This requirement that the states

Figure 6.16: ***Ring location on density plot of 1S and 2P States which are equally weighted***

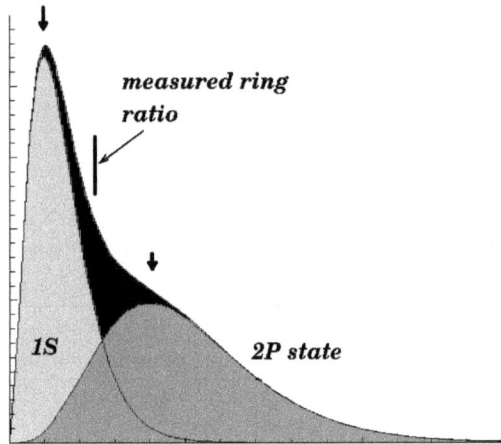

be large enough to support the necessary volume to acquire mass is easily stated in the next pair of equations. Where M is the mass of the central attractor and m is the mass of the captured particle.

$$g_n > a \quad when \quad M < \frac{\hbar n c}{Gm} \qquad (6.26)$$

This condition in equations 6.26 is easily met and only becomes a problem in the case of galaxies with masses greater than 10^{42} kg for individual minimum energy photons. For the case of the heavier photon for the sn1987a the threshold is 10^{35}kg which is also met for the stellar mass which is well below this threshold.

If the density of particles within the bound state is sufficiently high then a plane will exist on which the gravitational forces along the axis of the P states will be in balance from the central regions of the galaxy. In the case of sn1987a the ratio in ring radius is 2.33 well under the maximum allowed ratio of 3.33 for the 1S to 2P mean radius ratio. This indicates that there is significant fraction of the mass of the system in the 2P state. Matter collecting on the null F_z plane formed by the P state, can orbit and oscillate through the distributed mass of the bound particles. The position of this null plane can be computed as a function of the ratio of mass of the star and the mass of the P states.

Everything is actually set up to describe the rapid removal of energy from a massive hot burning stellar core. Three things are required: A very light boson $m < 10^{-50}$kg, a set of stationary states that are frozen and can take a ground state boson where the boson populations are growing, thirdly a very hot stellar core $\sim 10^9 K$ to efficiently drive the low energy boson radiation. If sn1987a stationary states are not formed by a new particle, or photons in BECs then only the kinetics of a single radiation processes must be understood. As the population of the bound ground state boson grows the transition rate is proportional to the number of bosons, $N(t)$, in that state so in time the occupation number can grow exponentially.

$$N(t) = Ae^{kt} \qquad (6.27)$$

This process will stop when the thermal energy of the stars interior has been quenched by the flow of bosons into the trapped state. The energy removal does two things. First, the transitions spread the energy upon transition over the much larger volume of the stationary state and it is essentially made transparent by being a weakly interacting state and not being able to interact with the plasma in the star. Second, once the thermal energy of the star has been significantly reduced its remaining mass will simply contract under its gravity driving the high energy nuclear fusion reactions that are responsible for the flash and the neutrino pulse of the supernova.

Summary

The sn1987a may give the first view of a new boson with a mass of 3.5×10^{-51} kg or a collection of bosons with this mass functioning as a single particle. A high rate of bound state boson production could be the mechanism that triggers a nova or supernova by quenching the core of a hot star and allowing it to collapse due to loss of internal thermal pressure.

Chapter 7

Comment on Quantum Mechanics

The arguments made here represent our understanding of the correspondence principal where there must be physical meaning in the context of the quantum dynamics of all properties brought over to classical mechanics. In the application of a model that encapsulates the many-body interaction of a particle with its medium, we have come up with three stable angular momentum states that can support bosons and at least one state that can support a fermion. Here, we are limited to two sets of quantum states with angular momentums of $(0,1,2)$ and $(0,2)$. These are the the only quantum states available that define mass as a quantum property and have a correspondence to mass for the representation of a classical particle. The property of mass must find meaning in both a macroscopic and microscopic description that is self consistent with the rules of quantum mechanics and relativity. Quantum mechanical description of bosons that interacts with the vacuum state is put together in a stacked hierarchy a little like thermodynamics relation to statistical mechanics. The break in properties for bosons occurs at the scale parameter, a, for each particle's many-body interaction with space in which it sits. Mass is an extrinsic property like a thermodynamic property, much like the specific heat of a substance. Mass loses its meaning for a boson examined below the scale of the many-body interactions with its space. As the scale approaches, a, in the space the particle occupies, the Schrödinger equation generates a solution that is a condition for mass in the case of the spherical flat bottom potential or whatever similar set of random potentials nature supplies for that interaction. For the boson relativity is handled conveniently in the upper range for the attractive potential and is not required for the massless lower range of activities.

These discussions started with the very low energy behavior of bosons, but the rules of quantum mechanics that govern them extend to other atomic matter by reversing the sign of the potential a particle can excite in vacuum defines a fermion. All particles with mass interact with the vacuum state. Particles that have no interaction with the vacuum state or potentials are massless at the scale they are observed. This picture is just an extension of a long evolving description of a particle in space that connects early works on gravitation and electromagnetic theory that was discussed by A. Einstein in his lecture at Leyden University, 5 May 1920 [73]. But preceding Einstein, Newton himself laid out the importance of this view and the method in the following quote:

"For many things lead me to have a suspicion that all phenomena may depend on certain forces by which particles of bodies, by causes not yet known, either are impelled toward one another and cohere in regular figures, or are repelled from one another and recede" [74]

Newton provides a tool and it is a powerful one, that can be applied on any scale. We have shown its application at a scale smaller than an atom and as large as a galaxy. We began with a self consistent solution to Schrödinger's equation that contained a massive particle, with the aim to probe the characteristics of the solution and find where the consistency of the solution breaks down for mass. For a solid state problem the many-body assumption is a reality and a difficult one from the point of view of doing accurate calculations. A particle in what we call a vacuum is a more fundamental assumption than an experimental reality, as we only see the collective results. Two interactions were considered, the first was an attractive weak interaction with the medium and this yielded a great deal of information about stable bosons. The second interaction was a repulsive interaction with a polarized medium that defines the conditions which mass could be found for a stable fermion. The first result from these assumptions is a new quantum wave equation that generates a solution for a particle with mass in its rest state. This new solution for a particle's density distribution yields a description that is invariant as a function of energy of the particle. This solution represents the quantization condition for mass. From that density distribution it is possible to derive the interaction potential directly required for the case of the fermion. This solution captures the basic random nature of the interaction process while allowing the infinities of the point particle interaction to be removed.

The form of the equation for mass in the case of the boson can be written $M \sim \kappa$ where κ is the curvature defining the attractive volume of the potential. For the fermion there was no simple solution, equation 3.11, except for the $\frac{1}{r}$ potential where mass shows the same dependence on curvature as the boson. From a physical point of view, the concept of mass not being an intrinsic particle property, but a characteristic that is acquired via a three-dimensional volume interaction, fits more closely with what is required to provide a connection between general relativity and quantum mechanics. The new wave equation captures the dynamics of both the fermions and the boson directly once mass is related to its source in the particles random interaction with vacuum or a dense medium. The new equation captures a series of events and organizes them into a description of what we call mass and a connection to what is required of mass in the general theory of relativity. The concept of time that actually can be measured is made more concrete by defining where it cannot be use as well as where special relativity no longer functions. This transition occurs when a particle is studied in its own frame of reference and requires a description suited to that particular frame of reference.

The very low mass of the pAxion which motivated this inquiry, is due to a weak magnetic interaction in steel over a large volume. When this concept of mass was extended to the photon, which has been searched for in numerous experiments, allowed for some very large scale quantum stationary states. An apparatus with a size on the order of the galaxy is required to measure the photon mass accurately. The photon rest energy changes how calculations will be performed because there is now a well-defined infrared cutoff frequency. The behavior of electromagnetic fields on the large scale also changes, with the concept of a static field being removed when the scale is great enough to have massive photons. The gravitational red shift that can be computed for an optical photon moving through vacuum with a mean density of $10^{-28} \frac{kg}{m^3}$ is too small to be detected. However, a photon originating in a higher density region moving through gravitational potentials will radiate which will contribute to a measurable red shift.

Photons captured in stationary states by large astronomical masses are a source of dark matter in the universe and a convenient mechanism to absorb rotational angular momentum from the mechanical spin down of matter. Studying the mass of something as light as 4.05×10^{-58}kg in vacuum requires the scale of astronomical observations. The Milky Way and one supernova show examples of the behavior that may represent stationary states on a large scale. New ringed structures[75] are being found in the microwave background radiation. These may be related to the gravitational effects of nearby or early stationary states. We expect there will be more examples of such activity that will be useful in defining the range of photon activity and dark matter on a large scale. The most important result in this book is the introduction of a procedure by which a new quantum wave equation can be derived from the concept of quantum Brownian motion to provide a description of the rest state of a free massive particle. This extends the solution spaces for quantum particles into a different domain and makes quantum mechanics consistent with Newton's first law. These new solutions not only illuminated the concepts of mass but also self energy, time and measurement.

Chapter 8

Appendix

8.1 Appendix A Repulsive Potential

Computing the eigenstates of the repulsive potential core problem when the core dimension is taken to very small values is equivalent to flipping the bound state problem over. The problem is defined as $V > 0$ for $r < A$ and $V = 0$ for $r \geq a$. Because the core is considered to shrink to small values, the same solutions used for the interior of the potential well problem can be used now on the exterior. Within the region of the high potential interior core region the solutions will have to be finite and decaying to the origin. The set of solutions on the interior are the hyperbolic function, rather than the sine-cosine function of the exterior set of solutions. These can be obtained by the variable substitution in the arguments of the signs and cosines by $x \to ix$.

$l = 0$ and $r < a$:

$$u(r) = A\frac{sinh(\gamma r)}{r} \tag{8.1}$$

$l = 1$ and $r < a$:

$$u(r) = A\left(\frac{sinh(\gamma r)}{r^2} - \frac{cosh(\gamma r)}{r}\right) \tag{8.2}$$

$l = 2$ and $r < a$:

$$u(r) = sinh(\gamma r)\left(\frac{1}{r} + \frac{3}{r^3}\right) - \frac{3cosh(\gamma r)}{r^2} \tag{8.3}$$

The function remains finite for $u(0)$ for $l > 2$ as $j_l(x) = \frac{x^l}{(2l+1)!!}$ [15] where $(2l+1)!! = (2l+1)(2l-1)\ldots.5 \times 3 \times 1$.

8.2 Appendix B - $F1(a, b, c)$ and $U(a, b, c)$ Hypergeometric Functions

The properties of U(a , b , r) can be expressed in the following terms [76] :

$$F1(a, b, z) = 1 + \frac{az}{b} + \frac{(a)_2 z^2}{(b)_2 2!} \ldots + \frac{(a)_n z^n}{(b)_n n!} + \ldots \tag{8.4}$$

where $(a)_o = 1$ and :

$$(a)_n = a(a+1)(a+2) \ldots (a+n-1) \tag{8.5}$$

Figure 8.1: *Two F1(a,b,c) examples of confluent hyperge-*
ometric based probability density function plotted. The
case when F1(2,2,kr) is plotted it is equal to a constant
1.

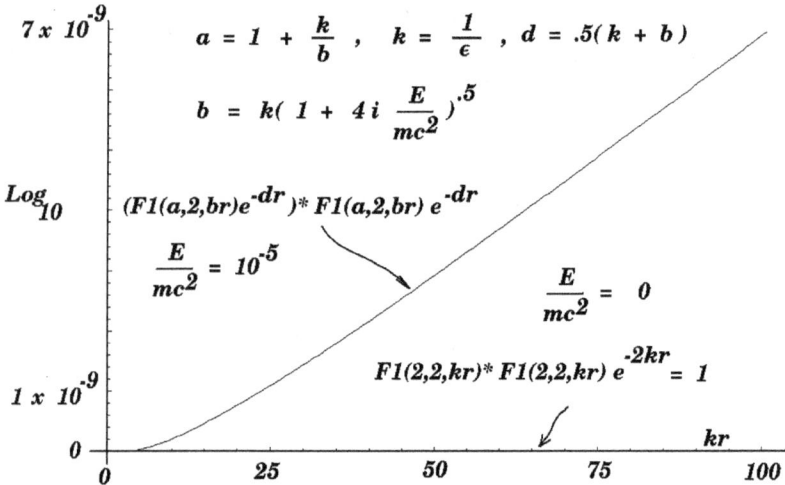

For the F1 based solutions, they are finite but not zero at the origin then at large amplitudes grow rapidly. Both the constant value and the rapid growth with radius make them unsuitable as solutions describing a particles interaction with the vacuum.

Where Γ represents the gamma function for positive integer arguments $\Gamma(n+1) = n!$.

$$U(a, b, z) = \frac{\pi}{sin\pi b} \{ \frac{F1(a, b, z)}{\Gamma(1+a-b)\Gamma(b)} - z^{1-b} \frac{F1(1+a-b, 2-b, z)}{\Gamma(a)\Gamma(2-b)} \} \tag{8.6}$$

$$U(2, 2, z) = \frac{\pi}{sin2\pi} \{ F1(2, 2, z) - z^{-1} \frac{F1(1, 0, z)}{\Gamma(0)} \} \tag{8.7}$$

Where

$$\lim_{z \to n} \frac{1}{\Gamma(-z)} = 0 \quad where \quad n = 0, 1, 2, 3 \ldots \tag{8.8}$$

For a ground state fermion in its frame of reference the wave function of the state can be approximated from the wave function derived.

$$u(r) = pe^{-\alpha r} U(2, 2, \alpha r) \quad for \quad E = 0 \tag{8.9}$$

The below function can represent an approximation of the above function in the range of κr being near one.

$$u(r) = A \frac{e^{-\frac{\kappa r}{2}}}{r^{\frac{3}{2}}} \tag{8.10}$$

where $A \sim \frac{1}{10^5}$. For the case $E \neq 0$ the function:

$$u(r) = pe^{-\frac{\kappa}{2}(1 + \sqrt{1 + 4i\frac{E}{mc^2}})r} \times U(1 + \frac{1}{\sqrt{\kappa^2 + 4i\frac{E}{mc^2}}}, 2, \sqrt{\kappa^2 + 4i\frac{E}{mc^2}}r) \tag{8.11}$$

The approximation becomes:

$$u(r) = A \frac{e^{-\frac{(1 + \sqrt{1 + 4i\frac{E}{mc^2}})\kappa r}{2}}}{r^{\frac{3}{2}}} \tag{8.12}$$

In the limits of very small r the approximation for U becomes, $U(2, 2, r) \simeq \frac{1}{r}$ and in the limits of very large r, $U(2, 2, r) \simeq \frac{1}{r^2}$
The expressions for the derivatives of the functions F1 and U:

$$\frac{d^n}{dz^n} F1(a, b, z) = \frac{(a)_n}{(b)_n} F1(a + n, b + n, z) \tag{8.13}$$

where:

$$F1'(a, b, z) = \frac{a}{b} F1(a + 1, b + 1, z) \tag{8.14}$$

For the U(a,b,z) function the derivatives are:

$$\frac{d^n}{dz^n} U(a, b, z) = (-1)^n (a)_n U(a + n, b + n, z) \tag{8.15}$$

and for the first derivative:

$$U'(a, b, z) = U(a, b, z) - U(a, b + 1, z) \tag{8.16}$$

or

Figure 8.2: *Comparison of the approximation function over the midrange of values for U confluent hypergeometric based wave function*

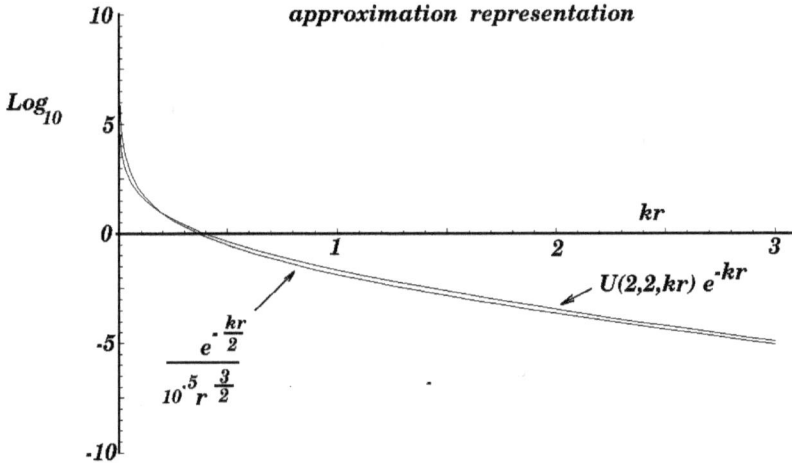

approximation representation

10

Log_{10} 5

0 kr

                              1 2 3

                                                $U(2,2,kr)\, e^{-kr}$

-5 $e^{-\frac{kr}{2}}$

          $\dfrac{}{10^{.5} r^{\frac{3}{2}}}$

-10

$$U'(a,b,z) = \frac{1-b}{z}U(a,b,z) - \frac{1+a-b}{z}U(a,b-1,z) \qquad (8.17)$$

The derivatives of U(a,b,z) taken in a sequence of higher derivatives appear as a self-similar set of functions which have been scaled.

Forth Code for 3 Particle Coulomb Interaction Using a Monte Carlo Integration

Source code for the Monte Carlo integration to produce the effective minimum relaxed potential experienced by a charged fermion is in the following listing. The minimum energy geometry is a linear arrangement of charge distributions with the positron in the center and the two electrons are moved outwards. The source code is in the Forth language [3] and is a dual stack language with an integer stack and a floating point stack. Operations are expressed in reversed Polish notations with operations on the stack elements. The code is organized into a linked list of Forth words which encapsulate subroutines which are referenced and executed from their address. This linked list of Forth words is called dictionary. A forth word begins with a colon and ends with a semicolon and blank space delineates all operators and symbols. If compact word construction is used the code can be made to be self-commenting.

Program File *repel_mc.f* **Comments in ( )**

include /usr/local/forth/lib/options/fpmath.f (floating point include)

(random number generator from Will Baden)

VARIABLE BUD

*: RAND (– u) BUD @ 3141592621 * 1+ DUP BUD ! ;*

: RND (n – u) RAND UM NIP ;*

: /RND (–)

COUNTER BUD ! ; (COUNTER is the current system timer count)

/RND

: rrnd 2000000 rnd 1000000 - s > f 1.0e-6 f ;*

: 3rrnd rrnd rrnd rrnd ;

*: FARRAY1Z CREATE 8 * ALLOT DOES > SWAP 8 * + ;*

fvariable wimple-r

fvariable wimple-dis

fvariable wimple1-density

fvariable wimple2-density

fvariable wimple3-density

3 farray1z wimple1-loc

3 farray1z offset 0e0 1 offset f! 0e0 2 offset f!

3 farray1z wimple2-loc

3 farray1z wimple2-org-loc

3 farray1z wimple3-loc

3 farray1z wimple3-org-loc

fvariable int-wimple

fvariable int-wimple+

fvariable int-wimple-

*: wimple function wimple-r f@ fnegate fexp wimple-r f@ 3e0 f** f/ ;*

: wimple-density

3 0 do i wimple1-loc f@ loop

fdup f fswap fdup f* f+ fswap fdup f* f+ fsqrt wimple-r f!*

wimple-function wimple1-density f!

3 0 do i wimple2-org-loc f@ loop

fdup f fswap fdup f* f+ fswap fdup f* f+ fsqrt wimple-r f!*

wimple-function wimple2-density f!

3 0 do i wimple3-org-loc f@ loop

fdup f fswap fdup f* f+ fswap fdup f* f+ fsqrt wimple-r f!*

wimple-function wimple3-density f! ;

: wimple-test fdup 1e-6 f < if fdrop 1e-6 then ;

*: wimple-potential wimple-density wimple1-density f@ wimple2-density f@ f**

0e0 3 0 do i wimple1-loc f@ i wimple2-loc f@ f- fdup f f+ loop fsqrt wimple-test f/ fnegate*

int-wimple- f@ f+ int-wimple- f!

*wimple2-density f@ wimple3-density f@ f**

0e0 3 0 do i wimple2-loc f@ i wimple3-loc f@ f- fdup f f+ loop fsqrt wimple-test f/*

int-wimple+ f@ f+ int-wimple+ f! int-wimple+ f@ int-wimple- f@ f+ int-wimple f! ;

```
: rwimple1 3rrnd 3 0 do i wimple1-loc f! loop ;
: rwimple2 3rrnd 3 0 do fdup i wimple2-org-loc f! i offset f@ f- i wimple2-loc f! loop ;
: rwimple3 3rrnd 3 0 do fdup i wimple3-org-loc f! i offset f@ f+ i wimple3-loc f! loop ;
( main word r-wim generates a table of potentials when executed )
: r-wim 200 0 do i s > f 0.1e0 f* 0 offset f! 0e0 int-wimple f! 0e0 int-wimple+ f! 0e0
int-wimple- f!
10000000 0 do rwimple1 rwimple2 rwimple3
wimple-potential
loop i 10 mod 0= if cr i 4 .r then
int-wimple f@ f > d 12 d.r
loop ;
```

Once there is a representation that is no longer a point charge, but a distribution that is as compact as the above density, the infinites that are generated by the point charge potentials vanish. If by chance the location selected for one interaction is the same for two different bodies that would generate an infinity, requires that element to be further subdivided and the random selection repeated for the two points. This can go on until they do not coincide on the scale of the iterated subdivision. This eliminates the infinite response. This procedure has not yet been included in the code above. In the code above if that case occurs then it is simply not computed, but can be estimated algebraically.

There is a question of whether the time for the disordered distribution to form is less than that of the measurement interval. Or more accurately put, can measurements be made on a time scale less than that required to generate the distribution that generates mass? For a boson that is not a problem because measurements can be made on a scale less than that required to define mass. With a reduced time scale, the wave functions and the potentials will be much steeper for the charged fermions.

If the fermion has mass then its activity with the vacuum state is intense, and the exchange interaction represents the difference between method of electron correlation. In the case of parallel spins, the correlation other than the individual particles interacting with the vacuum state in competitive manner is driven by simple repulsion. Whereas for anti-parallel spins, there is no competitive limitation on the interaction with the vacuum state. Those effects driven by the strong self potential dominate in creating the charge distribution and reduce the direct repulsion contributions.

8.2.1   The $\frac{1}{r}$ Potential

If a Coulomb like potential, $\frac{g}{r}$ is added to the new wave equation that results in the following expression:

$$u''(r) + (\frac{2}{r} + \kappa)u'(r) - \frac{i\kappa^2}{mc^2}(\frac{g}{r} + E)u(r) = 0 \qquad (8.18)$$

defining a set of parameters α and γ:

$$\alpha = \kappa^2 \frac{E}{mc^2} \qquad (8.19)$$

$$\gamma = 1 + \frac{\kappa + g}{\sqrt{\kappa^2 + 4i\alpha}} \qquad (8.20)$$

There is a valid solution that appears very similar to the potential free solutions, whereas, for the harmonic oscillator potential that does not occur:

$$u(r) = e^{-\frac{1}{2}(\kappa + \sqrt{\kappa^2 + 4i\alpha})r}$$
$$\times \{pU(\gamma, 2, \sqrt{\kappa^2 + 4i\alpha}r) + qF1(\gamma, 2, \sqrt{\kappa^2 + 4i\alpha}r)\} \qquad (8.21)$$

The effect of the $\frac{1}{r}$ potential is to scale the hyperbolic factors in the solutions. In fact the term $(1 + g)$ can be replaced by $(n - 1)$ where n is the dimensionality of the problem even for dimension of $n > 3$. These higher dimensional solutions should be useful in economic modeling.

8.3   Appendix C pAxion Experimental Methods

The pAxion dispersion experiment in 12 mm diameter hot rolled 1018 low carbon steel wire are described in reference [5]. The same procedure used in the scattering experiments in a steel wire describe in [5] were used for the scattering in the hot rolled .003 m thick 1018 carbon steel sheet. In both scattering experiments, drive level of the injected ω_2 signal is a variable that is increased until the single-frequency transitions are detected, i.e. $(\omega_1 \pm \omega_2)$ and $(\omega_1 \pm 2\omega_2)$. In the wire scattering experiment there was a significant displacement between the sources ω_1 and ω_2. That is not the case in the plate experiment.

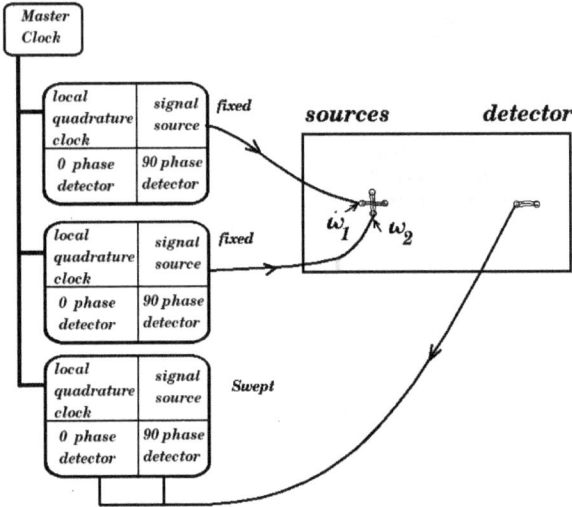

The sketch is a schematic of the experiment using either the Process Monitor 4 or 5 from Casting Analysis Corp. as a coupled set of signal sources and phase detectors which are phase locked on a master oscillator. Any of the phase detectors can be used as quadrature spectrum analyzer to allow the integrity of the measurement channel or the detector coil to be scanned for noise and signal content.

8.4   Appendix D Photon Eigenvalues

Photons in some circumstance may form and persist in a local Bose-Einstein condensation (BEC) since there is not a restriction against this even if the local temperatures exceed $10^9 K$.

The minimum required occupation number is less than one at even these extremely high temperatures because of the small mass of the photon. A BEC would have to be populated at the time of the photon creation into a state that would confine the BEC into a volume that would be close to the volume required for the photon to acquire mass. This may be possible when the size of the bound state is on the order of the radius required to acquire mass. Using a photon mass of 4.05×10^{-58} the gray regions are plotted which show where valid eigenvalues for bound states will be found as the binding potential well depth is increased for a radius determined by the rest state photon properties. On this graph there is plotted the minimum energy eigenvalue for the three lowest angular momentum states at -1.12×10^{-22} eV. There is another interesting feature in this plot and that is for the $l = 0$ case in the lower range as the potential well depth is increased.

Potential Well Depth Map of where photon eigenvalues are valid

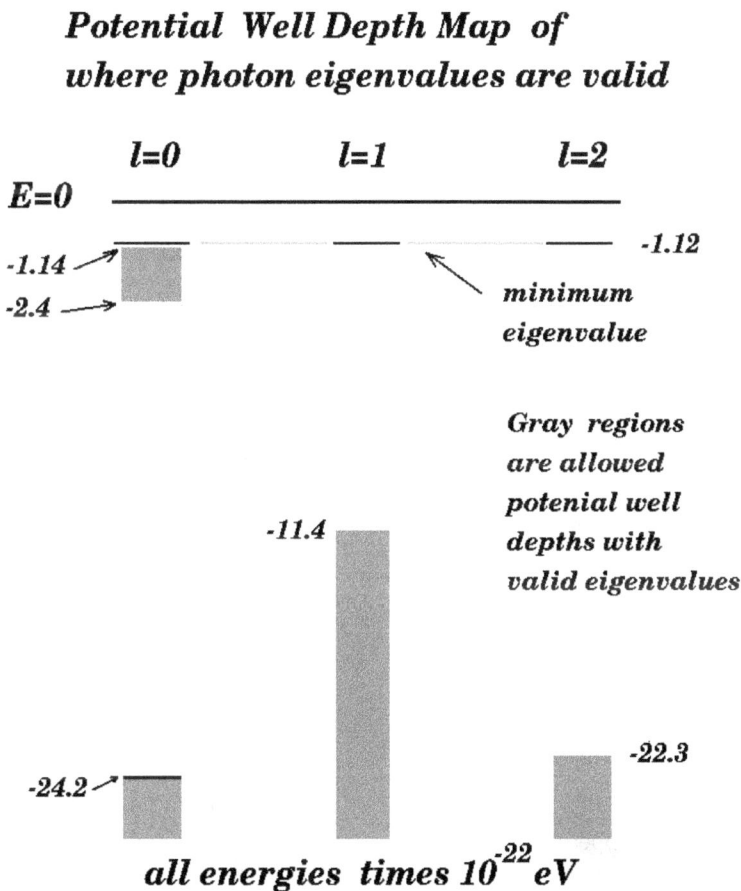

all energies times 10^{-22} eV

Here, a large and shallow potential well is modeled for the capture of photons. All values are in units of $10^{-22} eV$ where the radius of the well was determined by $a = \frac{\hbar}{mc}$ for a self energy of the photon at its minimum frequency in eV. These energies are considerably greater than the eigenvalues for the bound states of the Milky Way. The numerical eigenvalues are determined graphically from the transcendental equations that result from the continuity requirement. The $l = 1$ is in bold face for the photon's angular momentum.

The main feature for the $l = 1$ photon is that the next state that can be occupied requires a potential well depth is an order of magnitude greater than the rest energy. If such a well were constructed of the attractive interaction within a photon group at the lowest energy it would probably take a very large number to produce that potential. This would require a term analogous to a Gross-Pitaevskii attractive term in the modified Schrödinger equation used to describe Bose-Einstein condensates.

Table 13 Eigenvalues for a photon trapped in a minimum radius spherical potential.

V eV $\times 10^{-22}$	$l = 0$	$l = 1$	$l = 2$
- .5	-	-	-
- .7	-	-	-
- 1	-	-	-
-2	-.18	-	-
-3	-	-	-
-5	-	-	-
-7	-	-	-
-9	-	-	-
-10	-	-	-
-12	-	**-.3**	-
-14	-	**-1.5**	-
-16	-	**-2.8**	-
-18	-	**-4.3**	-
-20	-	**-5.8**	-
-22	-	**-7.6**	-
-24	-	**-9.2**	-.65
-26	-5.7	**-10.9**	-2.0

Just as a note you can compute the eigenvalues for angular momentums $l > 2$ and they appear at ever deeper potentials well depths. Much like the distribution of states for a bar rotator that has been used to motivate the necessity for string theory. The only problems with the distribution beyond $l = 2$ is that the eigenfunctions are not regular at the origin and would not represent a stable state. So a linear spectrum of excitations as a function of angular momentum may not require a string like model in order to generate an equivalent spectrum.

Reference Notes

The best references are usually the original papers because they will often be more clearly and precisely expressed. In particular the experimental papers are very important and those that deal directly with the analysis of experimental data such as classical Brownian motion work. In a written document it is difficult to capture the importance that laboratory and measurement experience brings to the understanding and analysis of data of all types. Mathematics in this book is relatively simple because the task is understanding the physics of mass and not classifying phenomenon. A general survey of the subject with a multitude of original sources is James R. Newman four volumes *The World of Mathematics* [77]. An elementary introduction to mathematics is R. Courant's and H. Robbins' *What is Mathematics?* [78] and for calculus there is the little book by Sylvainus P. Thompson *Calculus made easy* [79]. For a solid practical mathematical reference, we have used Kaplan's *Advanced Calculus* [80] for elementary concepts about the differential equation of mathematical physics and for complex analysis and their application to the equations used in doing physics there is *A Course of Modern Analysis* by Watson and Whittaker [31]. Even though the applications of group theory [81] are useful in some branches of quantum mechanics, its application when you are still discovering information on particular interactions can be seductively misleading when extended into areas not yet fully developed.

Real materials are our reference base and Alan Cottrell's *An Introduction to Metallurgy* [82] is an excellent description of matter spanning a range of scales. Atomic properties are cover in Max Born's *Atomic Physics* [83] and nuclear properties in E. Fermi's *Nuclear Physics* [59]. As previously mentioned the early editions of *Classical Mechanics* by H. Goldstein [29] gives a clear introduction of classical mechanics and special relativity. The clearest two basic texts on quantum mechanics that cover the subject from slightly different perspectives are by L. Schiff [25] and P.A.M. Dirac [30] which is necessary when learning the subject. There is a short concise set of notes on quantum mechanics by Enrico Fermi that maybe even better [84] because it makes the subject appear compact. For a well explained reference on electromagnetic theory, J.D. Jackson's *Classical Electrodynamics* 3rd edition [15] is very accurate and begins with a discussion of the history of the search for the rest mass of a photon. A recent book, *Quantum Field Theory* by Srednicki [43], was recommended and found to be extremely coherent in a subject that has many levels of possible confusion and makes a good companion to Heitler's *The Quantum Theory of Radiation* [13] that is explicit about defects in the theory. Unlike most subjects, mathematics and physics basic references are not subject to decay and it will be necessary for the physics to cover some points more than once to extract the full meaning.

There are very good histories written on most of the main subject areas and they are useful as they put in perspective the sequence and directions taken on these evolving subjects, particularly by Schweber [40] and Pais [85] [86]. There are a set of interviews on the nature of quantum mechanics that present some well developed points of views that are not found in the text books or histories but are in the words of the principals doing the work [87]. In hindsight, the development of quantum mechanics and its application probably had as much to do with the contributions of Wolfgang Pauli as any other figure because of the methods he used in quantization that avoided being engulfed in the details of specific many-body problems of atomic, molecular and nuclear physics. This formal application of group theory in a number areas along with the strong classical disposition to use the point mass concept by his contemporaries froze consideration of the details that give rise to the spin quantum, mass and the general description of the rest state. These topics are discussed by Pauli himself in a translation of some of his papers [45].

About the Authors

John P. Wallace attended Columbia University, the Henry Krumb School of Mines, and worked in the labs of Polykarp Kusch. He wrote theses for Martin Gutzwiller and John K. Tien. He worked for short periods at IBM Watson Labs, Magnetic Analysis Corp, Westinghouse Research Labs and University of Minnesota. Most of the last 30 years, he has been running his own company Casting Analysis Corp which develops and manufactures instruments used in crystal growth, metal defect detection and monitoring hydrogen in metals.

Michael J. Wallace attended Hampden-Sydney College and learned experimental physics from Weyland Joyner. He also built and operated ELF sounding tools with the help of Casting Analysis Corp instrumentation. He also worked at the NRAO in Greenbank, West Virginia on the calibration of the Byrd telescope, and later the analysis of VLBA data for Dr. Grant Denn formerly of Sweet Briar College. Attending Montana Tech he did a master's thesis for Prof. William Sill focused on Electrical Geophysics. Since graduating in 2006, he has been employed at Willowstick Technologies LLC doing ground water mapping as a geophysical service on dams, mines, oil fields, aqueducts and reclamation sites.

Bibliography

[1] J.P. Wallace and M. J. Wallace. Appendix c: paxion, photon and mass. In G.R. Myneni, G. Ciovati, and M. Stuart, editors, *SSTIN10 AIP Conference Proceedings 1352*, pages 313–335, Melville, NY, 2011. AIP.

[2] C.H. Moore and E.D. Rather. The forth program for spectral line observing on nrao's 36 ft telescope. *Astromomy and Astrophiscs Supplement Series*, 15(3), 1974.

[3] J.V. Noble. *Scientific Forth*. Mechum Banks Pub., Ivy Va., 1992.

[4] R.P. Feynman, R.B. Morinigo, and W.G. Wagner. *Feynman Lectures on Gravitation*. Westview Press, Bolder Co., 2003.

[5] J.P. Wallace. Electrodynamics in iron and steel, arxiv:0901.1631v2 [physics.gen-ph], 2009.

[6] J.P. Wallace. Spintronics enter the iron age. *JOM*, 61(6):67–71, June 2009.

[7] H. Koh and C.L. Magee. A functional approach for studying technological progress: Extension to energy technology. *Technological Forecasting and Social Change*, 75:735–758, 2008.

[8] A. Smith. *An Inquiry into the Nature and Causes of the Wealth of Nations*. Methuen and Co. Ltd., London, 5 edition, 1904.

[9] J.P. Wallace, G.R. Myneni, and R. Pike. Curvature, q and hydrogen. In G.R. Myneni, G. Ciovati, and M. Stuart, editors, *SSTIN10 AIP Conference Proceedings 1352*, pages 38–46, Melville, NY, 2011. AIP.

[10] M. Jammer. *Concepts of Mass in Classical and Modern Physics*. Harvard Univ. Press, Cambridge, Mass, 1961.

[11] G. Galilei. *Dialogues Concerning Two New Sciences*. Dover, NYC, 1954. trans. H. Crew, A. de Salvio.

[12] S. Chandrasekhar. *Newton's Principia for the Common Reader*. Clarendon Press, Oxford, 1995.

[13] W. Heitler. *The Quantum Theory of Radiation*. Oxford Univ. Press, London, 3rd edition, 1954.

[14] G. Hooft. Entangled quantum states in a local deterministic theory, arxiv:0908.3408v1 [quant-ph], 2009.

[15] J. D. Jackson. *Classical Electrodynamics*. John Wiley & Sons, NYC, 2nd edition, 1975.

[16] G. Feinberg. Possibility of faster-than-light particles. *Phys. Rev.*, 159:1089, 1967.

[17] J. Wallis. *De algebra tractatus*. 1st edition, 1673.

[18] M. Planck. *The Theory of Heat Radiation*. P. Blakiston's and Son, Philadelphia, 1914. trans. Masius, M.

[19] J.P. Wallace. Proton in srf niobium. In G.R. Myneni, G. Ciovati, and M. Stuart, editors, *SSTIN10 AIP Conference Proceedings 1352*, pages 205–312, Melville, NY, 2011. AIP.

[20] G. Wentzel. New aspects of the photon self-energy problem. *Phys.Rev.*, 74:1070–1075, 1948.

[21] J. Goldstone. Field thories with superconductor solutions. *Nuovo Cim.*, 19:15, 1961.

[22] Y. Nambu and G. Jona-Lasinio. Dynamical model of elementary particles based on an analogy with superconductivity. i. *Phys.Rev.*, 122:345–359, 1961.

[23] P. Anderson. Plasmons, gauge invariance and mass. *Phys.Rev.*, 130:439–442, 1963.

[24] G. Hooft. The glorious days of physics-renormalization of gauge theories, arxiv:9812203v2 [hep-th], 1999.

[25] L.I. Schiff. *Quantum Mechanics*. McGraw Hill, NYC, 2 edition, 1955.

[26] P. Nozières and D. Pines. Electron interactions in solids, general formulation. *Phys.Rev.*, 109:741–761, 1958.

[27] R. I. G. Hughes. Theoretical practice: the bohm-pines quartet. *Perspective on Science*, 4:457–524, 2006.

[28] R.D. Amado and J.V Noble. Efimov's effect: A new pathology of three-particle systems. ii. *Phys.Rev.D*, 5:1992–2002, 1972.

[29] H. Goldstein. *Classical Mechanics*. Addison-Wesley, Reading, Mass, 1950.

[30] P. A. M. Dirac. *The Principles of Quantum Mechanics*. Oxford Unv. Press, London, 4nd edition, 1958.

[31] E.T. Whittaker and G. N. Watson. *A Course of Modern Analysis*. Cambridge Univ. Press, Cambridge, 4nd edition, 1927.

[32] G. N. Watson. *A Treaties on the Theory of Bessel Functions*. Cambridge Univ. Press, Cambridge, 2nd edition, 1944.

[33] A. MacKey. *Mathematical Foundations of Quantum Mechanics*. W.A. Benjamin, Inc., NYC, 1963.

[34] E. Fermi. *Thermodynamics*. Dover Press, NYC, 1936.

[35] S. Raimes. *The Wave Mechanics of Electrons in Metals*. North-Holland Publishing Company, Amsterdam, 1961.

[36] A. Einstein. *Investigation on the Theory of the Brownian Movement*. Dover Press, NYC, 1926,1956. trans. A. D. Cowper, notes & ed. R. Fürth.

[37] M. Kac. Random walk and the theory of brownian motion. *American Mathematical Monthly*, 54(7):369–391, 1947.

[38] J. Schwinger. Brownian motion of a quantum oscillator. *J. Math. Phys.*, 2:407, 1961.

[39] M.C. Gutzwiller. *Chaos in Classical and Quantum Mechanics*. Springer-Verlag, NYC, 1990.

[40] S. S. Schweber. *QED and the Men Who Made It: Dyson, Feynman, Schwinger, and Tomonaga*. Princeton Unvi. Press, Princeton, N.J., 1994.

[41] M. Zralek. 50 years of neutrino physics, arxi:1012.2390v1 [hep-ph], 2010.

[42] R. Bellman. *Perturbation Techniques in Mathematics, Physics, and Engineering*. Dover, NYC, 1972.

[43] M. Srendnicki. *Quantum Field Theory*. Cambridge Univ. Press, Cambridge, 2007.

[44] B. Van Der Pol and H. Bremmer. *Operational Calculus*. CambridgeUniv. Press, NYC, 2nd edition, 1964.

[45] W. Pauli. *Writings on Physics and Philosophy*. Springer-Verlag, Heidelberg, 1994.

[46] A. Tarski. *Introduction to Logic*. Dover Press, NYC, 1995.

[47] T. Sharpe. *Vintage Stuff*. Martin, Stecker and Warburg Ltd., London, 1982.

[48] S. Weinberg. *The Quantum Theory of Fields Vol I and II*. Cambridge Unvi. Press, Cambridge, 1995.

[49] J. Baggott. *The Quantum Story*. Oxford Unvi. Press, Oxford, 2011.

[50] D. P. DiVincenzo and E. J. Mele. Self-consistent effective mass theory for intralayer screening in graphite intercalation compounds. *Phys.Rev.B*, 295(4):1685, 1984.

[51] Z.Q. Li, E.A. Henriksen, Z. Jiang, Z. Hao, M.C. Martin, P. Kim, H.L. Stormer, and D.N. Basov. Dirac charge dynamics in graphene by infrared spectroscopy. *Nature Physcis*, 4:532–535, 2008.

[52] J.D. Bjorken and S.D. Drell. *Relativistic Quantum Mechanics*. McGraw-Hill, NYC, 1964.

[53] M. Su, T.R. Slatyer, and V.P. Finkbeiner. Giant gamma-ray bubbles from fermi-lat: Agn activity or bipolar galactic wind?, arxiv:1005.5480v3 [astro-ph.he], 2010.

[54] G.R. Denn, R.L. Mutel, and A.P. Marscher. Very long baseline polarimetry of bl lacertae. *Astrophysical J. Supp. Series.*, 129:61, 2000.

[55] S. Gillessen, F. Eisenhauer, S. Trippe, T. Alexander, R. Genzel, F. Martins, and T. Ott. Monitoring stellar orbits around the massive black hole in the galactic center, arxiv:0810.4674v1 [astro-ph], 2008.

[56] G. Feinberg. Pulsar test of a variation of the speed of light with frequency. *Science*, 166(3907):879–881, 1969.

[57] O.K. Baker, G. Cantatore, J. Jaeckel, and G. Mueller. Notes from the 3rd axion strategy meeting, arxiv: 1007:1835v1[hep-ph], 2010.

[58] H.A. Bethe and E.S. Salpeter. *Quantum Mechanics of One- and Two-Electron Atoms*. Springer, Berlin, 1957.

[59] E. Fermi. *Nuclear Physics*. University of Chicago Press, Chicago, Ill., 1949.

[60] E. R. Williams, J. E. Faller, and H. A. Hill. New experimental test of coulomb's law: A laboratory upper limit on the photon rest mass. *PRL*, 26:721, 1971.

[61] A. S. Goldhaberand and M. M. Nieto. Terrestrial and extraterrestrial limits on the photon mass. *Rev. Mod. Phys.*, 43:277, 1971.

[62] E. Fischbach, H. Kloor, and R. A. Langel. New geomagnetic limits on the photon mass and on long-range forces coexisting with electromagnetism. *PRL*, 73:514, 1994.

[63] R. Lakes. Experimental limits on the photon mass and cosmic magnetic vector potential. *PRL*, 80:1826, 1998.

[64] J. Luo, L. C. Tu, Z. K. Hu, and E. J. Luan. New experimental limit on the photon rest mass with a rotating torsion balance. *PRL*, 90:081801, 2003.

[65] F. Zwicky. Die rotverschiebung von extragalaktischen nebeln. *Helvetica Physica Acta*, 6:110, 1933.

[66] F. Zwicky. On the masses of nebulae and clusters of nebulae. *Astrophysical J.*, 86:217, 1937.

[67] L. Chemin, C. Carignan, and T. Foster. H i kinematics and dynamics of messier 31. *Astrophysical Journal*, 705:1395, 2009.

[68] H. Shapley. *Galaxies*. Harvard Univ. Press, Cambridge,Mass, 3rd edition, 1972.

[69] J.S. Lundee, B. Sutherland, A. Patel, C. Stewart, and C. Bamber. Direct measurment of the quantum wave function. *Nature*, 474:188–191, 2011.

[70] R. McCray. Supernova 1987a at age 20. In S. Immler and K. Weiler, editors, *Supernova 1987A: 20 Years After, AIP Conference Proceedings 937*, pages 3–14, Melville, NY, 2007. AIP.

[71] V.P. Utrobin. Supernova 1987a: ejecta mass and explosion energy. In S. Immler and K. Weiler, editors, *Supernova 1987A: 20 Years After, AIP Conference Proceedings 937*, pages 25–32, Melville, NY, 2007. AIP.

[72] H.A. Bethe. Supernova mechanisms. *Rev. Mod. Phys.*, 62(4):801–866, Oct. 1990.

[73] A. Einstein. *The Collected Papers of Albert Einstein, Vol 7*. Princeton Unvi. Press, Princeton, N.J., 2002. trans. A. Engel.

[74] I. Newton. *Newton: Philosophical Writings*. Cambridge Univ. Press, Cambridge, 2004.

[75] V.G. Gurzadyan and R. Penrose. Concentric circles in wmap data may provide evidence of violent pre-big-bang activity, arxiv:1011.3706v1 [astro-ph.co], 2010.

[76] L. J. Slater. Confluent hypergeometric functions. In M. Abramowitz and I.A. Stegun, editors, *Handbook of Mathematical Functions with Formulas, Graphs, and Mathematical Tables*, ASM 55, pages 503–536. Dept. of Commerce, Washington DC, 1968.

[77] J.R. Newman. *The World of Mathematics*. Simon and Schuster, NYC, 1956.

[78] R. Courant and H. Robbins. *What is Mathematics*. Oxford Univ. Press, NYC, 2nd revised by i. stewart edition, 1996.

[79] S.P. Thompson. *Calculus Made Easy*. St. Martin's Press, NYC, 3rd edition, 1986.

[80] W. Kaplan. *Advanced Calculus*. Addison-Wesley, Reading, Mass., 1952.

[81] H. Weyl. *The Theory of Groups and Quantum Mechancis*. Dover Press, NYC, 1932. trans. H. P. Robertson.

[82] A. Cottrell. *An Introduction to Metallurgy*. Edward Arnold Ltd., London, 1967.

[83] M. Born. *Atomic Physics*. Dover Press, NY, 1969.

[84] E. Fermi. *notes on Quantum Mechanics*. University of Chicago Press, Chicago, Ill., 1961.

[85] A. Pais. *Subtle is the Lord*. Oxford Univ. Press, NYC, 1982.

[86] A. Pais. *Inward Bound*. Oxford Univ. Press, NYC, 1986.

[87] P.C.W. Davies and J.R. Brown, editors. *The Ghost in the Atom*. Cambridge Univ. Press, Cambridge, 1986.

Index